# INTRODUCTION

## Ghosts in the Machine

The human labor powering many mobile phone apps, websites, and artificial intelligence systems can be hard to see—in fact, it's often intentionally hidden. We call this opaque world of employment *ghost work*.[1] Think about the last time you searched for something on the web. Maybe you were looking for a trending news topic, an update on your favorite team, or fresh celebrity gossip. Ever wonder why the images and links that the search engine returned didn't contain adult content or completely random results? After all, every business, illicit or legitimate, advertising online would love to have its site ranked higher in your web search. Or think about the last time you scrolled through your Facebook, Instagram, or Twitter feed. How do those sites enforce their no-graphic-violence and no-hate-speech policies? On the internet, anyone can say anything, and, given the chance, people certainly will. So how do we get such a sanitized view? The answer is people and software working together to deliver seemingly automated services to customers like you and me.

Beyond some basic decisions, today's artificial intelligence can't function without humans in the loop. Whether it's delivering a relevant newsfeed or carrying out a complicated texted-in pizza order, when the artificial intelligence (AI) trips up or can't finish the job, thousands of businesses call on people to quietly complete the project. This new digital assembly line aggregates the collective input of distributed workers, ships pieces of projects rather than products, and operates across a host of economic sectors at all times of the day and night. In fact, the rise of this shadow workforce is part of a larger, more profound reorganization of employment itself. This yet-to-be-classified form of employment done on demand is neither inherently good nor bad. But left without definition and veiled from consumers who benefit from it, these jobs can easily slip into ghost work.

Businesses can collect projects from thousands of workers, paid by the task. Now they can depend on internet access, cloud computing, sophisticated databases, and the engineering technique of human computation — people working in concert with AIs — to loop humans into completing projects that are otherwise beyond the ability of software alone. This fusion of code and human smarts is growing fast. According to the Pew Research Center's 2016 report *Gig Work, Online Selling and Home Sharing,* roughly 20 million U.S. adults earned money completing tasks distributed on demand the previous year.[2] Professional, white-collar information service work, delivered through on-demand work platforms, is already projected to add $2.7 trillion, or 2.0 percent, to global GDP by 2025.[3] If trends continue at the current rate, economists estimate that by the early 2030s, tech innovation could dismantle and semi-automate roughly 38 percent of jobs in the U.S. alone.[4] Left unchecked, the combination of ghost work's opaque employment practices and the shibboleth of an all-powerful artificial intelligence could render the labor of hundreds of millions of people invisible.

Who does this kind of work? People like Joan and Kala.

Joan works from the Houston home she shares with her 81-year-old mother. In 2012, Joan moved in to care for her mother after a knee surgery left her mom too frail to live on her own. A year later, Joan started picking up work online through MTurk — short for "Amazon Mechanical Turk," a sprawling marketplace owned and operated by tech giant Amazon.com. Joan makes some of her best money doing "dollars for dick pics." That's how she describes labeling pictures flagged as "offensive" by social media users on platforms like Twitter and Match.com.

Companies can't automatically process every piece of content users

# CONTENTS

# Ghost Work

# GHOST WORK

## How to Stop Silicon Valley from Building a New Global Underclass

**Mary L. Gray and Siddharth Suri**

*Houghton Mifflin Harcourt*
Boston New York
2019

hmhco.com

*Library of Congress Cataloging-in-Publication Data*
Names: Gray, Mary L., author. | Suri, Siddharth, author.
Title: Ghost work : how to stop Silicon Valley from building a new global underclass / Mary L. Gray and Siddharth Suri.
Description: Boston : Houghton Mifflin Harcourt, 2019. | Includes bibliographical references and index.
Identifiers: LCCN 2018042557 (print) | LCCN 2018044155 (ebook) | ISBN 9781328566287 (ebook) | ISBN 9781328566249 (hardcover) | ISBN 9780358120575 (int. ed.)
Subjects: LCSH: Labor supply — Effect of automation on. | Automation — Economic aspects. | Artificial intelligence — Economic aspects. | Technological unemployment.
Classification: LCC HD6331 (ebook) | LCC HD6331 .G826 2019 (print) | DDC 331.1 — dc23
LC record available at https://lccn.loc.gov/2018042557

Book design by Chloe Foster

Printed in the United States of America
DOC 10 9 8 7 6 5 4 3 2 1

For Ila and George

—M.L.G.

To my family and in loving memory of my dad

—S.S.

flag for review, so some of the harder-to-evaluate materials are routed to workers like Joan. On the surface, her task seems simple: click on pictures and assess their content. Is that an X-rated penis selfie that should be removed, or some innocuous G-rated body part? She is paid for each task she completes and decides when she walks away from her computer. Joan, with years of practice, now knows how to piece together an average ten-hour day that will bring in roughly $40* worth of such tasks.

Thousands of miles away in Bangalore, India, Kala works from her makeshift home office, tucked away in the corner of her bedroom.[5] Joan and Kala do similar tasks, sorting and tagging words and images for internet companies, but Kala picks up work from an outsourcing company that supplies staff to the Universal Human Relevance System (UHRS), an MTurk-like platform used internally by its builder, Microsoft. Kala, a 43-year-old housewife and mother of two with a bachelor's degree in electrical engineering, calls her two teenage sons into the room, points to a word displayed inside a large text box on her LED monitor, and asks them, "Do you know what this word means? Is it something you shouldn't say?" They giggle as she reads the text out loud to them. They make fun of her pronunciation of "chick flick." Together they decide that, no, this sentence does not contain adult content. Kala clicks "no" on the screen, and the window refreshes with a new text phrase to read to her sons. "They are more qualified to recognize these words than me," she says, laughing. "They help me keep the internet clean and safe for other families." Though she's typically unable to find enough tasks to fill more than 15 hours of work in a given week, Kala returns to UHRS almost every day to see if there are any new tasks that she feels qualified to do. Kala's doggedness and luck in the past have paid off. Now that she's learned how to browse and claim tasks quickly, Kala can make the time she has between making meals and checking her children's homework feel, as she puts it, "fruitful" as she does web research for what she considers extra income.

* Throughout this book, all dollar amounts are in U.S. dollars. Rupee amounts refer to the Indian national currency in current conversion rates.

Content moderation — from sifting through newsfeeds and search results to adjudicating disputes over appropriate content to help technology and media companies figure out what to leave up or take down — is just one example of a new type of work that depends on people like Joan and Kala. Reviewing content is a common, often time-sensitive task generated in the wake of social media companies' attempts to identify family-friendly materials for the billions of people who use their sites every day. There are way too many webpages, photos, and tweets in every imaginable language for people like Joan and Kala to assess them all.

Companies like Google, Microsoft, Facebook, and Twitter use software to automatically remove as much "not safe for work" content as they can, wherever possible. But these software filtering systems, powered by machine learning and artificial intelligence, aren't perfect. They can't always tell the difference between a thumb and a penis, let alone hate speech and sarcasm. Remember that classic moment in the 2012 U.S. presidential campaign when Republican candidate Mitt Romney uttered the phrase "binders full of women!"? Twitter needed workers, doing the same type of work that Joan does, to figure out, in real time, why a hashtag attached to such an obtuse phrase was quickly soaring to the top of its trending topics. Was it a hack? A glitch? Bona fide, frenetic Twitter use? Current AI systems can't reliably tell the difference. On-demand work offers the promise of blending the power of computation with the creativity and dynamism of human insight.

This book is the story of Joan, Kala, and the millions of workers like them who step in when AI falls short. They are the humans behind the seemingly automated systems that we all take for granted. But modern AI systems don't just need humans to answer an unfamiliar or challenging question; they also need humans to help them learn how to answer anything in the first place. For example, do an image search for "camelback couch" and you'll get a whole bunch of pictures of couches with curved backs. Search engines like Bing and Google don't see or understand images in the way we humans do. Furniture aficionados need no more than a second to recognize a swank piece of furniture with a curved back that multiple people can sit on as a camelback couch. The AI systems behind search engines must start with at least a few hundred images of curve-backed couches, each labeled "camelback couch." Then, when the search engine encounters a new picture of a couch, it runs what is called a "classification algorithm," which essentially checks to see if the couch in this new image matches the geometrical patterns of those labeled "camelback" more than those not labeled "camelback." Now, where did the initial set of labeled images, called training data, come from? From people like Justin. With no more than a two-sentence task description as guidance, workers like Justin must claim a job within seconds or lose it to someone else willing to scoop up the job first. Justin's a stay-at-home dad with two young sons, working around his kids' preschool and nap schedules. He readily admits he had no idea what a camelback couch was at the start. "I had to spend an enormous amount of time on Google trying to look up these terms to figure out what they meant before I could answer the questions."

TripAdvisor, Match.com, Google, Twitter, Facebook, and Microsoft are some of the better-known businesses that generate an array of projects that people like Justin are paid to do, task by task, 24 hours a day, seven days a week. New companies crop up every day with business models that depend on workers around the world who respond to open calls routed through software to do this behind-the-scenes work. Businesses that can contract out their day-to-day activities to independent workers instead of regular employees can use ghost work to answer a web based customer chat query, edit a product review, or do just about any task that doesn't require an employee's full-time, physical presence.

## How Does Ghost Work Work?

A computer program is no more than a list of instructions that tell a computer what to do. When two software programs (or a piece of software and a piece of hardware) need to communicate, they must first establish a common language. They do so via an application programming interface, or API. The API determines the common language by defining the list of instructions that a program will accept and what will happen after each instruction is executed. One could say that the API specifies the computer program's "rules of engagement." For example, there are hundreds if not thousands of different kinds of computers on the market right now, so writing a custom version of a software system for each type would be impossibly complex. But when all (or at least significant fractions) of the machines available obey the same API, programmers can write code once for all of these kinds of machines, because the API ensures that all of the machines understand the same language. These types of APIs are limited to what a computer can do, but the MTurk API enabled software developers to write programs, using only a slightly different set of instructions, that automatically pay *humans* to do tasks.[6]

Normally, when a programmer wants to compute something, they interact with a CPU through an API defined by an operating system. But when a programmer uses ghost work to complete a task, they interact with a person working with them through the on-demand labor platform's API.[7] The programmer issues a task to a human and relies on the person's creative capacity—and availability—to answer the call. Unlike CPUs, humans have agency: they make their own decisions. While CPUs just execute whatever instruction they are given, humans make spontaneous, creative decisions and bring their own interpretations to the mix. And they have needs, motivations, and biases beyond the moment of engagement with the API. Given the same input, a CPU will always output the same thing. On the other hand, if you send a hungry human into a grocery store, he or she will walk out with a dramatically different bag of groceries than if they were not hungry. In exchange for this impetuousness and spontaneity, humans bring something to work that CPUs lack: creativity and innovation. Joan, Kala, and Justin are members of a growing economy, hidden by APIs and fueled by ghost work.

Less than two decades ago, software developers only wrote code for computers to execute. The MTurk API, and those that followed, allowed programmers to use humans to do tasks that are beyond a computer's capacity, like accurately making a quick judgment call, as Kala and Joan do when they determine what is and isn't adult content. In fact, anyone sitting in front of a web browser could now answer an automated request for help. Businesses call this mix of APIs, rote computation, and human ingenuity "crowdsourcing," "microwork," or "crowdwork." Computer scientists call it "human computation." Any project that can be broken down into a series of discrete tasks can be solved using human computation. Software can use these APIs to manage the workflow and process the output of computers and individuals and even pay people for their contributions once they have completed the task. These people power modern AI systems, websites, and apps that we all use and take for granted.

Imagine a woman in her early twenties — let's call her Emily — standing on a curb in Chicago. Emily opens the Uber app on her smartphone and an Uber driver responds. Neither Emily nor the driver knows that their meeting hinges on another woman, two oceans away — perhaps her name is Ayesha.[8]

Emily and her driver have no idea that Uber's software just flagged his account. The driver — let's say his name is Sam — shaved off his beard last night for his girlfriend's birthday. Now the selfie he took this morning — part of Uber's Real-Time ID Check, rolled out in 2016 to authenticate drivers — doesn't match his photo ID on record. It didn't occur to Sam that a discrepancy between the two photos — one showing him with a beard, one without — would automatically suspend his account. But suddenly, and unbeknownst to him, his livelihood hangs in the balance.

Meanwhile, overseas in Hyderabad, the Silicon Valley of India, Ayesha sits at her kitchen table, squinting at her laptop. She just accepted a job routed from Uber to CrowdFlower's software, and now she is an invisible yet integral part of the ride. CrowdFlower and its competitors with similarly hip-techy names, like CloudFactory, Playment, and Clickworker, offer their platform's software as a service to anyone who needs quick access to a ready crowd of workers. Tens of thousands of people like Ayesha log on to crowdsourcing platforms like CrowdFlower every day, looking for task-based work. Now Ayesha — and any other invisible workers who happen to have responded to CrowdFlower's request — will determine whether Sam picks up Emily.

Uber and CrowdFlower are two links in a growing supply chain of services that use APIs and human computation to put people to work. Uber uses CrowdFlower's API to pay someone to review the results of Ayesha's work, and, if it passes muster, it will process Uber's payment to her within minutes. If it doesn't meet the preprogrammed bar, Ayesha won't get paid for her efforts, nor will she have any meaningful opportunity to lodge a complaint. The API isn't designed to listen to Ayesha.

Ayesha compares the two photos of the driver side by side. A timer in the top right-hand corner of CrowdFlower's webpage winds down, prompting her to speed up. If she doesn't submit a response before the timer runs out, CrowdFlower won't process Uber's payment for the task. Ayesha blinks, glances at the timer, and squints at the thumbnail-size photos: *Yes, those are the same brown eyes. The same dimpled cheeks.* She clicks "okay."

Sam's account is authorized to pick up Emily just as he pulls up to the curb. Emily stops scanning the congested Chicago traffic and climbs into his car. By the time the car door closes, Ayesha has moved on to the next task. She hopes to net a few more rupees before she ends her workday.

Neither Uber's passengers nor their drivers realize that a person, working far away or perhaps just down the road, might vet their transaction in real time. Imperceptible exchanges like this one determine one out of every 100 Uber pickups in the United States, which means they happen roughly 13,000 times a day. We never saw the ghost work that Ayesha could do for CrowdFlower, but, having spent time with her and workers like her, we can imagine the fleeting market exchanges that consumers like Emily and drivers like Sam will never see. Ayesha is the only artifact of ghost work's presence and, as such, the only one who can help us recover the experience of ghost work after Emily and Sam are long gone.

• • •

Billions of people consume website content, search engine queries, tweets, posts, and mobile-app-enabled services every day. They assume that their purchases are made possible by the magic of technology alone. But, in reality, they are being served by an international staff, quietly laboring in the background. These jobs, dominated by freelance and contingent work arrangements rather than full-time or even hourly wage positions, have no established, legal status. Sometimes these jobs are given heft as harbingers of the "Second Machine Age" or the "Fourth Industrial Revolution" or part of a larger digital or platform economy. Other times, they're simply, glibly called *gigs*.[9]

No employment laws capture the on-demand gig economy's odd mix of independence from any single employer and dependency on a web-based platform. As the taskmasters of the gig economy, on-demand platforms make their money by matching those buying and selling human labor online, generating a two-sided market of myriad businesses and anonymous crowds of workers. And, importantly, as media scholar and sociologist Tarleton Gillespie points out, platforms may not create the content that they host, "but they do make important choices about it."[10] On-demand work platforms can easily become silent business partners more aligned with the interests of those willing to pay a fee to find workers than with the workers searching for jobs.

From the largest firms to the smallest startups, companies rely on this shared pool of on-demand workers amassed by on-demand platforms. They use this assembly of workers to satisfy customers who have grown to expect responses to their requests within seconds. Businesses turn to this pool, instead of traditional temporary staffing agencies, to fill last-minute gaps on their teams. They draw from it to spin up new projects, from testing a new software privacy setting to vetting descriptions of culturally attuned mac-and-cheese flavors. Such ventures are too speculative or loosely understood to justify hiring a full-time employee or the expense of recruiting, even through a temp service. No business wants to invest in launching a new service or product without gauging how consumers will respond. Service industries, driven by the ever-shifting winds of customer taste and satisfaction, can try out ideas generated by ghost work and iterate on responses from other workers, standing in for the average consumer.

## Robots Might Be Coming, but They Aren't Here Yet

Every week, another breathless headline proclaims the end of work. Soon, we are warned, the robots will rise up against us. Automation and its handmaiden, artificial intelligence, are widely understood as processes making human labor obsolete. Robotic arms can move sheets of metal across the factory floor. Software bots can take texted pizza orders. Drones can deliver packages to our doorsteps. These intelligent systems, now hitched to many traditional employment sites, are said to herald the rapid disappearance of humans in the workplace. The inevitable triumph of AI, so the story goes, will make all but the most uniquely qualified workers redundant. We all need to skill up. *Now.*

Tesla and SpaceX founder Elon Musk, renowned physicist Stephen Hawking, and Google co-founder Larry Page are just a few of the prominent voices in this chorus.[11] Either they express panic about "summoning the demon" of AI or wax nostalgic about a time before AI, when humans supposedly controlled their own destiny.[12] But arresting headlines obscure a messier reality. While it's undeniably true that robots are on the rise, most automated jobs still require humans to work around the clock, often part-time or on a contract basis, fine-tuning and caring for automated processes when the machines get stuck or break down, as technical systems, like humans, are apt to do.

It's also true that the long march toward automation has historically created new needs and different types of human labor to fill those needs. In this respect, the new, software-managed work world shares features of the factory jobs that assembled cars by placing workers on a production line where and when they were needed most. It also resembles the so-called piecework that women and children did on farms in the 19th century, assembling matchstick boxes for pennies a pop. And it overlaps in obvious ways with the outsourcing of medical transcription and call center work to the Global South that boomed with the expansion of the internet in the late 1990s.

Factory work, piecework, and outsourcing were all precursors to tasks distributed online insofar as they involved jobs that were small, repetitive, and removed from the bigger picture. These jobs came with little stability or support. They were done, most often, by people whom economists might consider expendable or "low skill." The market calls this, unironically, "human capital." Clicking "dog" or "cat" to label an image that will eventually enable an iPhone to recognize a family pet is not that different from turning a screw on what will eventually become a Ford truck. But that's where the job similarities end.

Blue-collar manufacturing jobs have been the most visible targets of AI's advance. The Foxconn factories that make iPhones allegedly replaced 60,000 humans with robots in 2016. Amazon's 20 fulfillment centers reportedly deployed 45,000 robots to work alongside 230,000 people that same year. Yet these numbers confound how many jobs are *created* by automation. And the media coverage of AI's impact on full-time blue-collar work can distract us from the rapid growth of a new category of human workers to complement or tend to automated manufacturing systems when AI hits its limits.

In the past 20 years, the most profitable companies have slowly transitioned from ones that mass-manufacture durable goods, like furniture and clothing, to businesses that sell services, like healthcare, consumer analytics, and retail. There's more money to be made in selling consumers an experience, from sipping a latte to watching a bit of infotainment, than building a television set.[13] Businesses of all types manage costs by tapping into and maintaining control of a pool of contingent workers.

Having who you want, when you want them, is now a half-century-old strategy for avoiding negotiations with full-time employees and the classification and employment laws that protect them.

This hybrid of humans and AI reconfiguring manufacturing, retail, marketing, and customer service has outstripped familiar employment categories. Unlike the repetitious lockstep of factory-controlled, full-time manufacturing shift work, these task-based services, such as correctly amending a client's tax return or translating and captioning a video in real time, depend on endless iterations of human discernment and divination that don't fit neatly into a traditional 40-hour workweek. The tasks are dynamic, not merely mechanical, which is why it is difficult to eliminate humans from the task at hand.

AI is simply not as smart as most people hope or fear. Take, for example, the celebrated accomplishments of the AI powering AlphaGo, most recently chronicled in technologist Scott Hartley's book *The Fuzzy and the Techie.*[14] In May 2017, AlphaGo became the first computer program to beat Ke Jie, the reigning world champion of the ancient Chinese board game go. Five months later, AlphaGo fell to its progeny, AlphaGo Zero. But, lest we be too impressed, it's important to keep in mind that the rules of go are fixed and fully formalized and it is played in a closed environment where only the two players' actions determine the outcome. AlphaGo and AlphaGo Zero's human programmers at the Google-backed company DeepMind gave the programs clear definitions of winning versus losing. Winning go is about foreseeing the long-term consequences of one's actions as one plays them out against those of an opponent.[15] So AlphaGo was trained on billions of board positions using a large database of games between human experts, as well as games against itself, allowing it to learn what constitutes a better move or a stronger board position.[16] AlphaGo Zero was then steeped in all of those prior experiences by playing against AlphaGo, a mirror image of self. But, as Tom Dietterich, a noted expert in artificial intelligence research, suggests, "we must rely on humans to backfill with their broad knowledge of the world" to accomplish most day-to-day tasks. Real life is more complicated than a game of go.

The new online work platforms that channel jobs to Joan, Kala, Justin, and Ayesha upend the mediagenic stories about AI's boundless wisdom and the inexorable rise of robots. Real-world tasks, from identifying hate speech or categorizing a rental as a great springtime wedding venue to correctly amending a tax return, require human discernment. Formalizing the singular, best choice, as you might in a game of go, won't work. For example, it would be difficult, if not impossible, to enumerate every attribute of a wedding venue that would make it the "best." Even if this were possible, people would have different preferences when it came to the attributes of the venue. Moreover, the training data to teach AI to recognize what counts as the "best choice" does not exist. In addition, an endless set of external factors, from vernacular slang and climate-change-induced hurricanes to haphazard tax reform legislation, can intrude and influence the outcome. In many cases, there are too many unknowns to train current AIs to be aware enough or gain enough experience to intelligently respond to all cases of the unexpected. This is why AI must return to humans to backfill decision-making with their broad knowledge of the world.

Anyone who scrutinizes the shadows of AI, as we have done, will find a new world of work in which software manages people doing jobs that computers can't do. As builders create systems to transfer tasks from humans to machines, they surface new problems to solve through automation. For example, it was only after the web became mainstream that companies like Facebook, Twitter, and Instagram faced growing demand to moderate their online content, outstripping the limited capacity of automated moderation tools. At the same time, as novel systems are brought online, they typically face unanticipated problems and fall short of their promise, hence the need for Kala's and Joan's work. Thanks to workers like them, automated moderation software is better, but it is far from perfect. The inevitable glitches that automated processes encounter along the way to perfection generate temporary work for people. Once they have successfully trained artificial intelligence to perform like humans, workers move on to the next tasks engineers assign them that push the boundaries of automation. Since the finish line moves as people dream of new applications for AI, we can't be sure if the "last mile" of the journey toward full automation will ever be completed. We call this the "paradox of automation's last mile."

As AI advances, it creates temporary labor markets for unforeseen and unpredictable types of tasks.[17] The great paradox of automation is that the desire to *eliminate* human labor always *generates* new tasks for humans. What we call "the last mile" is *the gap between what a person can do and what a computer can do.* Without a doubt, software developers will use ghost work to perform the tasks at hand and push AI to its limits. And it is just as likely that as more companies aspire to give us AI-enabled "smart" digital assistants to manage our calendars and book our flights, we'll need more and more people to step in when AI falls short of our increasingly exacting and extensive demands. In fact, dependency on temporary human labor has always been a part of the history of technology's long march toward automation. Today's engineers aiming to solve problems through algorithms and AI are the latest iteration of the paradox of automation's last mile. On this frontier, the peaks and valleys of temporary work shift constantly, redefining relationships between humans and machines in the process.

The rise of on-demand labor platforms signals the allure of using APIs to organize, route, and schedule work. As the examples in this book suggest, this reorientation to use contingent labor to develop new technologies fueled the recent "AI revolution." When an AI system that powers a phone app or online service isn't confident about what to do next for a customer, it needs human help, and it needs it fast. End users expect software running search engines and social media to respond in milliseconds. Traditional methods of hiring won't do here. So if an AI needs a human in the loop, to make sense of a spike in search terms tied to, say, a sudden natural disaster, it needs to get human input immediately.

The disaster will fade into history. The software will have learned what it needed from the momentary flood of human input. That is exactly what an always-on labor pool, plugged into APIs, provides. Software developers can write code that automatically hires someone to solve an immediate problem, checks their work, and pays them for doing the job. Similarly, scientists and researchers using modern machine learning systems depend on training data that's clear and error-free. They need an automated method to get help generating and cleaning up that data, and they rely on many people around the world to do it. On-demand labor platforms offer today's online businesses a combination of human labor and AI, creating a massive, hidden pool of people available for ghost work. Delivering services and jobs on demand could be an integral part of the future of work. It could also have unintended, potentially disastrous consequences if not designed and managed with care and attention to how it is restructuring the experience and meaning that people attach to their day jobs.

## Ghost Work and the Future of Employment

The dismantling of employment is a deep, fundamental transformation of the nature of work. Traditional full-time employment is no longer the rule in the United States. It used to be that a worker could spend decades showing up day after day to the same office, building a career, with the expectation of getting steady pay, healthcare, sick leave, and retirement benefits in return. Now, centuries of global reforms, from child labor laws to workplace safety guidelines, are being unraveled. In fact, according to the U.S. Department of Labor's Bureau of Labor Statistics, only 52 percent of today's employers sponsor workplace benefits of any kind. In the wake of the Great Recession, Americans have come to realize that the best alternatives to serving food, providing healthcare, or selling goods in brick-and-mortar shops are the growing number of jobs that can be found in the on-demand gig economy. Because this work doesn't fit any ready-made classification in employment law, the terms-of-service agreements for platforms like MTurk and CrowdFlower are almost indistinguishable from the boilerplate dialogue boxes that we all click to update our software, erasing the protections that traditional workers enjoy.

While the Pew Center's best estimate puts the number of individuals involved in ghost work today at around 20 million, there is no corroborating tally of how many people like Joan, Kala, Justin, and Ayesha cobble together contract-based ghost work gigs to make ends meet. When the Bureau of Labor Statistics added a supplemental survey of Contingent and Alternative Employment Arrangements to the U.S. Census Bureau's May 2017 Current Population Survey (CPS), a monthly snapshot of 60,000 eligible households that provides the nation's employment and unemployment data for the U.S. Bureau of Labor Statistics (BLS), it was the first time it had tried to gauge the growth of contingent jobs in more than a decade.[18] According to the BLS's estimates, 10.1 percent of U.S. workers work without an explicit or implicit long-term employment contract. But this survey counts only people who hold an alternative employment arrangement as their primary or stand-alone job. So if a person does ghost work while also holding down a nine-to-five job with a single employer for a set salary or hourly wage — a very common trend among the most active workers we met — they are even harder to identify, let alone count.

The Bureau of Labor Statistics' 2017 Contingent and Alternative Employment Arrangements supplement to the Current Population Survey poses two hurdles for measuring the rise of ghost work. It is hard to really understand what "long-term employment" means to workers in a multiple-choice survey. It might be as hard to know what "primary job" means when so many people hold down multiple jobs to make their rent. The confusion over how to think about old work categories, like "long-term" or "primary job," is reflected in a head count from the Government Accountability Office that diverges with the BLS's numbers. It reported, just two years earlier, that at least 31 percent of the U.S. workforce claims that it does some form of alternative work arrangement that includes freelancing or independent contract work for hire.[19] Labor economists

Lawrence Katz and Alan Krueger estimate that temporary and alternative contract-driven work delivered through self-employed workers or those temporarily employed by staffing agencies — the so-called casualization of the workforce — rose from 10 to 16 percent, accounting for all net employment growth in the U.S. economy in the past decade.[20] The closest we might come to understanding the size and growth of ghost work comes from independent think tanks rather than governmental data.

The most conservative estimates of on-demand gig labor markets come from the Economic Policy Institute. Economist Lawrence Mishel and his research team estimate that between 0.5 and 1 percent of working adults in the U.S., or 1.25 to 2.5 million people, participate in the on-demand gig economy. But they come to that number through a very specific study of Uber drivers and the assumption that Uber and other ride-hailing mobile apps make up the bulk of gig work. A study produced by the JPMorgan Chase Institute found that 4.3 percent of U.S. adults, or 10.73 million people, had worked an online-platform-economy job at least once between 2015 and 2016.[21] A revolving door of temporary tasks defines this job market. No obvious professional title. No ladder. No bonuses. No guarantees. Tasks are finite, built to disappear once a firm has reached its specific target and the people hired to hit it have moved on to other projects.

From software engineering and legal services to commercial media and healthcare, a wide range of businesses now turn to on-demand labor platforms to convert white-collar careers into bundles of projects. Such all-digital information services and knowledge work convert the creative expertise required to think with and massage data into the consumable services delivered online by industries from tech and law to finance and entertainment. Because of these seismic shifts, the days of large enterprises with full-time employees working on-site are numbered. A crowded field of companies compete to sell information services that pair computers and smart devices with artificial intelligence. Companies like Catalant (formerly HourlyNerd), Popexpert, and Upwork use APIs to deliver the larger "macro-tasks" of knowledge work, on demand, to other businesses or individuals. The future of employment wrought by automation will undoubtedly be far more disjointed than traditional nine-to-five work. Some labor economists argue that a new reality of "fissured workplaces" is the ultimate result of turning long-term employment into a series of short-term contracts throughout the 1980s and '90s.[22] And yet this newly unpredictable reality hasn't dissuaded millions of digital workers around the world from sitting down at their keyboards day and night and performing the countless behind-the-scenes tasks that make our apps seem smarter than they are. This means that the future of business and employment will more likely resemble today's on-demand economy than a dystopian sci-fi film in which humans disappear and robots rule. It will require people to navigate layers of software interfaces and learn to labor in the shadow of AI. It will contain an ecosystem of independent contractors like Joan, typing away in spare bedrooms, cafés, and cinder-block homes in rural India, Knoxville, Tennessee, and Portland, Oregon—or anywhere else a person with an internet connection, a computer, ambition, or financial need can get online. When little attention is paid to the workers behind these jobs, on-demand labor can quickly become alienating, debasing, precarious, and isolating ghost work.

All of the workers we interviewed have something unexpected in common: hope. They *hope* to use on-demand jobs to control when they work, who they work with, and what tasks they take on. They *hope* to stay close to their families. They *hope* to avoid long commutes and hostile work environments. And they *hope* to gain experience that refreshes their résumé or opens a door to new possibilities. Also true is that many saw few other options for themselves or their families. Full-time employment in their towns often meant an hourly wage at a big-box store, working a fixed shift, adapting to unpredictable work schedules, and without meaningful opportunities to advance. On-demand jobs gave them real-world experiences scheduling meetings, testing and debugging websites, developing computer expertise, finding sales leads, and managing full-time employees' HR files. What worker doesn't hope to one day fully control both the schedule and the purpose of their workdays?

*Ghost Work* draws on a five-year study in which we—an anthropologist and a computer scientist and the research team we mustered—investigated this booming yet still largely hidden sector of the economy.[23] It is the culmination of more than 200 interviews and tens of thousands of survey responses collected from workers across the United States and India; dozens of behavioral experiments and social network analyses of on-demand work platforms; and unique studies of this labor market's other key players, namely the people turning platforms into businesses and those hiring workers on them. It exposes a world in which steady work and salaries are being replaced by a chaotic string of small projects and micropayments, and human bosses are being replaced by automated processes that are programmed to oversee a far-flung workforce of anonymous independent contractors. *Ghost Work* departs from the well-known story about the rise of robots by documenting a more complicated future that is already emerging. It shows how ghost work platforms foster our belief in the magical promise of technology.

As an anthropologist, Mary had her interest sparked by the specter of an atomized world of workers earning money by sorting and annotating thousands of pictures of pointy-eared dogs, hairless cats, and "dick pics." When Mary asked those hiring workers what they knew about the people picking up their tasks, the responses ranged from "I don't know" to "Why would I want to know that!?" As a computer scientist, Siddharth had used on-demand platforms for years to conduct online behavioral experiments, but he knew little about the workers, as the API kept them hidden from him.[24] Who were the people offering themselves up for hire? What motivated them to do what many consider "mindless tasks," and how did they make this ill-defined form of employment pay off? What did this work mean to them? How many tasks flow online through these on-demand platforms? What are the business models that produce the demand for task-based work? What are the overall workings of this task-based economy?

When our research team started asking these questions in 2013, the only people in the conversation were economists, computer scientists, and businesspeople. All three groups evaluated the on-demand labor market on the basis of its ability to enhance efficiency and maximize a company's bottom line. When humans did happen to come up in the discussion, it was in reference to the consumer. What was the quality of the consumer's experience? The engineers and computer scientists building APIs, for companies or for their own experiments to advance AI, wanted to design systems that eliminated what they assumed were costly, superfluous operations that annoyed end users. They were in the business of building smarter, faster software that could automatically match people to services, whether it was a ride, a meal, or tax advice, with an end goal of using the data from each iteration to train future software to automate even more. Few people were tracking what this approach to productivity would mean for the people who vied to do task-based work for hire. They operated from the assumption that the workers needed to generate training data and improve software would disappear once the AI got things right. Companies were building software, after all, not temp jobs.

For the next five years, we did something our respective research fields had not: we learned about the range of ghost work and the lives of people doing it by conducting one of the most comprehensive studies of its kind. *Ghost Work* is the first book to illuminate ghost work's role in building artificial intelligence and the lives of workers who are invisible yet central to the functioning of the internet and the future of automation. It offers an intimate, detailed look at the experience of workers in this new economy. We focus on workers living in India and the United States, the two countries with the largest on-demand labor pools, both with a long, entwined history of technological advancement. Our team interviewed and observed hundreds of people, in their homes and other makeshift workspaces, as they did everything from flag tweets to transcribe doctors' visits. We surveyed thousands more to establish a baseline to help us gauge which practices were typical and which were exceptional. We then scaled up the findings from our interview data by conducting dozens of behavioral experiments and "big data"–style analyses, each with thousands of participants. Throughout *Ghost Work,* the reader will see us toggle between these two types of analysis, combining their strengths to shed more light on those who work in the on-demand economy.

We examined four different ghost work platforms: Amazon.com's Mechanical Turk (MTurk); Microsoft's internal Universal Human Relevance System (UHRS); the socially minded startup LeadGenius; and Amara.org, a nonprofit site dedicated to translating and captioning content for transnational audiences and people with hearing disabilities. Each of these four platforms offers different products and business models. Investigating them alongside one another helped show us that our observations and conclusions hold broadly across the on-demand economy, as opposed to being specific to one category of ghost work. MTurk, as one of the first commercially available ghost work platforms, set the norms for how others would apply human computation to business solutions. UHRS stands in for the internal platforms that every large tech company maintains to meet its own ghost work demands. LeadGenius and Amara illustrate just how complex and sophisticated ghost work can be, as well as how much companies can play a role in designing better conditions for ghost work.

And then there were the workers. Among those working on these platforms, we met people stringing together on-demand projects to re-create the work hours, pay rates, and career development associated with full-time employment. We also met college-educated, stay-at-home parents staving off boredom; first-generation college students working 50 hours a week to save money for a wedding or fund a younger sibling's degree; and people, disabled or retired, looking for alternative routes to employment or extra money to pad their social security checks. We also met engineers and entrepreneurs who founded, designed, and built ghost work platforms.

When we started, we wondered: Who are these people, and how does their work differ from traditional nine-to-five jobs? On many on-demand labor platforms, a requester like Siddharth sees no personal information about a worker—gender, location, age, and prior work experience are all unknown. And workers have no information about the requester beyond the task description. The range of tasks can be endless and can change from one day to the next. APIs can be used to have a human tag a cat photo or run a research experiment, and similar APIs can be used to hire someone to deliver a meal, send a car, or design a website. The moment that the API is called and the work is produced looks automated to both consumers and requesters. But who benefits from this veneer of automation? And who might be harmed?

By the time we finished our study, we understood that people doing ghost work were no different from our friends and family making a living through freelance writing, research, software development, or adjunct teaching. Their work lives were often vulnerable and insecure. Yet the anonymity and remote access of on-demand platforms also made it easier for those marginalized in formal employment — because of where they lived, a perceived disability, or their belonging to a stigmatized minority — to earn an income.

The more closely we looked at the nascent edges of on-demand work, the more we saw people using familiar strategies to stay afloat and create meaningful employment for themselves and their peers. Sometimes these workers succeed by collaborating with one another. They share strategies for making difficult tasks easier, they swap intel about those with tasks for sale, and they help one another stay awake as they wait for new tasks to come online. We met workers who learned to move forward after their failed forays. Who learned to thwart exploitative business models, labor laws, and APIs designed to be indifferent to their interests. And we noted that businesses have no clue how much they profit from the presence of workers' networks. This book describes the thoughtless processing of human effort through APIs as *algorithmic cruelty* — literally, computation incapable of thought, let alone empathy. People doing ghost work understand the perils and potential of on-demand work better than any engineer, tech company CEO, policy maker, or labor advocate. They live it every day. And they are the most invested — economically and psychologically — in making it better.

Just as we need companies to be accountable for the labor practices that produce our food, clothes, and computers, so should the producers of digital content be accountable to their consumers and workers. We should demand truth in advertising in cases where humans have been brought in to benefit us — whether it is to curate our news or field complaints about what some troll just posted to our favorite social media site.

Along with a call for transparency, *Ghost Work* holds lessons for tech entrepreneurs who want a productive workforce, engineers who are building the labor platforms of the future, and policy makers charged with shaping this new commercial landscape. But the still untold story of the invisible workers who power the apps on our phones and the websites we look at should interest a wide range of general readers who've seen some coverage of "gigging it" or "Turk work," not to mention "crowdsourcing" and "microwork," and heard *a lot* about the rise of robots but want a deeper look at how, exactly, AI reshapes the working world and what, precisely, people do in the shadow of it. We offer a textured, nuanced, and ultimately hopeful account. Among other things, we show how moving beyond the full-time-freelance divide alone could go a long way toward sharing the wealth generated by the internet with those tasked to grapple with the paradox of automation's last mile. We hope, too, that the lessons we learned from the many workers we interviewed in the U.S. and India will help the millions of people who already, or will soon, do this work make the most of it. More than anything, *Ghost Work* is for anyone who works and wants to see what their future holds.

# PART I

## The Paradox of Automation's Last Mile

# 1

# Humans in the Loop

## The Beginning of Ghost Work

In the early 2000s, Amazon.com was in a tight spot. A young startup at the time, it had a rapidly expanding marketplace of books just as "e-commerce" was taking off. To build out the online bookstall, Amazon electronically pulled data about millions of book titles directly from publishers' catalogs and library listings. Much of it was rife with incorrect database entries. To build a loyal customer base, the company desperately needed to locate listing duplicates, typos, and outdated dust-jacket images linked to new editions before it could sell them to a still cautious segment of consumers willing to buy products online.[1] At first, Amazon hired temporary workers to clean up its databases. Following in the footsteps of neighboring tech companies, Amazon hired temps in both the U.S. and India, where it could get native English language fluency at low pay rates through staffing agencies that hired and managed contract workers. Not only did workers correct book titles, publication dates, and descriptions; they also made sure book covers matched the editions listed, and they embedded keywords in each book's web page, which were designed to return as many relevant items as possible when customers used Amazon's search bar.[2]

Then, in order to become the largest online retailer, Amazon would need to conquer a related challenge. Rather than stock millions of products, the company populated its site with small and medium-size businesses willing to list their inventory, from electronics and toys to cleaning supplies and niche foods, and sell their products through the Amazon site.[3] As Amazon grew its online marketplace to include other booksellers and then other products beyond books, it faced the additional challenge of ensuring that every description of a product matched its accompanying image. All of these merchants, including Amazon, would need a small standing army of workers willing to handle these discrete, repetitive tasks — like verifying that product descriptions matched the photo and creating captions and keywords to help online shoppers browse the ever-growing catalog of goods. Amazon turned to its vendors to hire contract workers to fill the labor demand. And as Amazon.com grew, it also needed to polish customers' book reviews so that awkward or unclear wording and syntax didn't muddy the value of customers' comments on its site. Soon contract workers did that work, too.

In 2005, Amazon.com publicly debuted a website it had built to make it easier for anyone with a verified account to clean up product listings and customers' typo-ridden reviews. The company called it Amazon Mechanical Turk, a name that users quickly shortened to MTurk. MTurk was an online labor market where "requesters" could post various tasks they needed done and workers could do those tasks for pay. The platform would list tasks and pay rates, as easily as you might post a job on Craigslist. The platform also operated as a bank so that those with work could fill up their payment accounts and keep a tab that could be used to automatically pay workers once they turned in their projects.

Amazon added a percentage surcharge to whatever the requesters paid workers for each task, charging requesters extra for matching them with workers guaranteed to have certain qualifications. Anyone willing to share their bank account, credit card information, and verifiable mailing address with Amazon could sign up to work on Amazon Mechanical Turk and earn credit toward gift cards on the virtual superstore. Payouts for tasks ran anywhere from a penny, for adding keywords to a specific image, to $25, for doing a single marketing survey. It's rumored that MTurk was a pet project of Jeff Bezos himself. Amazon's founder is said to have created MTurk so that Amazon could not just offer a marketplace of books and other durable goods but make labor itself a service that anyone could find and pay for through the Amazon website.[4]

In its first two years — perhaps a signal that the global recession was beginning to build — more than 100,000 people created accounts to find work on the MTurk platform. MTurk automated the hiring and paying of these workers, allowing any company or individual with a set of tasks piling up to issue them to a crowd that was logged on and looking for work. Not long after the platform picked up business beyond Amazon's own product teams, MTurk ironed out the logistics of direct deposit for cash payments to workers with mailing addresses in the U.S. and, for workers based in India, cutting paper paychecks that converted U.S. dollars to Indian rupees, routed through the Bank of Singapore. Once MTurk could pay out in cash instead of gift cards, a vast pool of U.S. and Indian workers vying for posted jobs were pitted against other workers who were more casually topping off Amazon gift cards from other countries.[5]

MTurk filled a much-needed niche. Companies and individuals putting more and more of their products online needed some way to check the accuracy of their posted materials. Increasing numbers of people responsible for entering their receipts into expense reports for reimbursement at work could now turn to services like MTurk for quick help from a human able to take on the task, any time in the day, any day of the week. Startups at the time, like Yelp and those contractors hired to write and curate content for their databases, could offer accurate restaurant locations for neighborhoods that had never had such detailed listings before. Marketing and PR agencies could distribute short surveys to get hundreds of reactions to new product ideas, slogans, and word associations for less than the hourly pay of a full-time employee or temporary worker. Academics could now survey a broader population, sending out 1,000-person polls in a matter of an hour, with results comparable to what they might get from distributing a similar poll to an introductory undergraduate course.

Even better, MTurk's surveys were more likely to reach a broader age and geographical demographic than the 18- to 22-year-olds found on most U.S. college campuses. While popular online classified listings like Craigslist in the United States had always included ads for work that people could do online, MTurk represented something entirely different. The platform offered contract work for tasks that required very little qualification or advanced computer experience. One needed time, attention to detail, and an internet connection. Whether for marketing, surveying, generating training data, or reviewing online content in real time, results from on-demand labor market platforms like MTurk could be obtained faster and more cheaply than anything collected from employees in the office. Soon, myriad new businesses formed, looking to profit from this mix of simple computer programming, a web interface, and unregulated hiring practices, producing a powerful new way to automate recruiting humans for ghost work.

Ghost work fueled a revolution in artificial intelligence through millions of people carrying out billions of small tasks hidden from technology consumers. Other companies soon figured out how to use ghost work to complete larger work projects, which we call "macro-tasks." In either case, ghost work powers many popular websites and mobile phone apps today while keeping the workers hidden behind the APIs used to hire them.

## Using APIs to Hire People

Anyone with access to MTurk's application programming interface (API) could plug into MTurk's growing pool of registered workers. Software developers, whether Amazon's own or those working for another company, could now write software to place tasks on the MTurk platform, making it easy to hire workers, evaluate their work, collect their projects, and pay them — all within a matter of seconds.

Programmers had previously only written code for machines to execute. But MTurk's innovation allowed humans to execute part of the programmer's code, as opposed to having only machines executing the code.[6] The brilliant breakthrough of MTurk was that it took this basic technology—batching a series of tasks so that they could be completed by humans through an API—and turned it into a labor market where people could "buy" and "sell" human labor. Now the same piece of software could simultaneously draw on human creativity and a computer's ability to crank through the same or similar tasks over and over. A programmer's software and MTurk's APIs effectively operated as managers of an on-demand contingent workforce. In the process, APIs and web-based platform interfaces, replicating MTurk's business model, seemed to eliminate much of what many of us expect from our bosses—feedback, scheduling, workspace, payment, and affirmation that we've done a proper job and completed our tasks. In doing so, MTurk unraveled the role of "employers" and turned programmers and companies seeking immediate help with tasks into "requesters."

The foremost implication of this new way of doing work is that the API determines the dialogue and communication between the programmer and the worker. For example, the API gives each individual requester and worker their own unique identifier, a string of seemingly random letters and numbers such as "A16HE9ETNPNONN." From the programmer's perspective, it makes the humans seem interchangeable, as each worker is represented by a worker ID, and everything that makes a human a person, such as their beliefs, attributes, and experiences, is stripped away from this identifier.

Computer scientists would say that all of your attributes are *abstracted* away. It's as if you were hired by someone who knew only your social security number and absolutely nothing else about you. The abstraction resulting from the API makes it appear as if there is no need to figure out who the humans are. The same way that poker chips can make gamblers forget that they are gambling with actual money, representing people as unique identifiers can make programmers forget that they are hiring *people* and their code affects people's lives. Amazon's somewhat callous reference to the 19th-century "Mechanical Turk" chess automaton wasn't as arcane as it might seem.[7] Amazon meant to draw parallels between its service and the enigmatic parlor game that toured for more than 80 years after its creation, in 1770. Ironically, the Mechanical Turk turned out to be a hoax, no more than a series of expert, small-statured chess players hiding inside the machine's wooden case. Humans in the loop — not machines — were the masterminds behind the automaton's chess moves. And, as the name suggests, it takes human intelligence to push the boundary of what machines can learn. APIs are the perfect taskmasters for teaching machines to advance AI.

## Ghost Work, Machine Learning, and the Rise of AI

Computer scientist Kevin P. Murphy defines machine learning as "a set of methods that can automatically detect patterns in data, and then use the uncovered patterns to predict future data."[8]

Recall the machine learning problem, from the introduction, of recognizing a camelback couch. A common machine learning approach would be to first gather what is called training data, in this case by gathering images of couches, say from furniture catalogs and social media posts, and having humans like Justin label them as "camelback" or "not camelback." Then the machine learning algorithm would compare a new image of a couch against images in the training data. If it looks more like a camelback couch, the algorithm would classify the new image as such. Now say the lighting in the new image is bad, or the angle doesn't show its back clearly, or there are people sitting on the couch and obscuring its back, so that the machine learning algorithm might not know how to classify it. This is where yet more human help might come in.

### ASSEMBLING IMAGENET

The overall goal of AI is to build computer systems with intelligence — the ability to evaluate and to act — comparable to what you might expect from another human being. Understanding what objects are in an image is a part of the ambitious revolution reaching for general artificial intelligence. After all, even a one- or two-year-old can recognize an apple or a dog in an image. Fei-Fei Li, a computer science professor and co-director of the Stanford Human-Centered AI Institute, and her colleagues wanted to solve a much more general problem than how to train AIs to recognize a specific object, like a couch. They wanted to train machines to recognize the main object in an image, no matter what that object may be—a dog, a person, a car, or a mountain. To do it, they needed more training data than a single person could generate alone. *Much* more.

Li and her colleagues first wrote software to download millions of images from the World Wide Web. At first, they hired a team of undergraduates to label each image—the academic equivalent of hiring a temp worker. After testing this method, they could extrapolate how long it would take to complete—about 19 years. So they switched strategies. Next they tried developing machine learning algorithms to automatically guess labels for images and turn to human help for the ones that stumped the machines. This approach failed, because the machine learning algorithms made too many errors, and they were looking for highly accurate or "gold standard" data that other scientists could later reuse. Indeed, if this problem were easily solvable by machines, they wouldn't have needed this data set in the first place.

Shortly after, in 2007, Li and her colleagues found MTurk, and they realized that the MTurk API gave them a way to automatically distribute image-labeling tasks to people and pay them. They tried a few different workflows but were ultimately able to use about 49,000 workers from 167 countries to accurately label 3.2 million images.[9] After two and a half years, their collective labor created a massive, gold-standard data set of high-resolution images, each with highly accurate labels of the objects in the image. Li called it ImageNet. Thanks to ImageNet competitions held annually since its creation, research teams use the data set to develop more sophisticated image recognition algorithms and to advance the state of the art. Having a gold-standard data set allowed researchers to measure the accuracy of their new algorithms and to compare their algorithms with the current state of the art. This allowed researchers to make so much progress that some AIs can now do a better job than humans in recognizing images![10]

The algorithmic and engineering advances that scientists achieved during the competition, between 2010 and 2017, fueled the recent "AI revolution," which had an impact across a variety of fields and a variety of problem domains. The size and quality of the training data were vital to this endeavor. MTurk workers are the AI revolution's unsung heroes. Without them generating and improving the size and quality of the training data, ImageNet would not exist.[11] ImageNet's success is a noteworthy example of the paradox of automation's last mile in action. Humans trained an AI, only to have the AI ultimately take over the task entirely. Researchers could then open up even harder problems. For example, after the ImageNet challenge finished, researchers turned their attention to finding *where* an object is in an image or video. These problems needed yet more training data, generating another wave of ghost work. But ImageNet is merely one of many examples of how computer programmers and business entrepreneurs use ghost work to create training data to develop better artificial intelligence.[12]

## The Range of Ghost Work: From Micro-Tasks to Macro-Tasks

The platforms generating on-demand ghost work offer themselves up as gatekeepers helping employers-turned-requesters tackle problems that need a bit of human intelligence. Businesses no longer had to turn to contingent staffing agencies to access a global labor market. MTurk became known for "micro-tasks," like the ones that Fei-Fei Li's team generated, that could be done quickly but required many people. But a host of businesses have cropped up in recent years that match workers to larger projects called "macro-tasks." These platforms, like Upwork and Fiverr, where you can find people to copyedit a newsletter, develop webpages, or build a mobile app, use the same employment strategy: distribute tasks to a pool of internet-connected workers who are hired, scheduled, managed, and paid, at least in part, by AIs and APIs. All those paid to complete tasks do a variation of platform-based ghost work. And, as of today, everyone is operating outside the legally defined classifications we have for employment. That is, there are no laws governing who counts as an "employer" or "employee" in ghost work. And it is unclear where the platforms, which are the places the workers go to find work, stand. But it is clear that the platforms have become the de facto job sites of on-demand labor. It's very difficult to see what it looks like to do this work until you meet the people behind the APIs.

## MTURK: THE PUBLIC FACE OF MICRO-TASKS

Joan, whom we met in the introduction, wears her hair in a loose bun skewered by shiny black chopsticks to keep it out of her eyes while she's working. She's been living in Houston since 2011, when she returned to care for her 81-year-old mother. Joan cooks, keeps house, and drives her mother to doctor appointments. And for the past three years, she's made most of her income working on Amazon's Mechanical Turk.

Before moving back to her hometown, Joan had a full-time job as a technical writer. She drafted and copyedited, among other things, manuals for filing for unemployment insurance in the state of Texas. At first, Joan lived off the money she'd cashed out of her 401(k) plan. But as her mom's health worsened, Joan looked for work she could do from home. On-demand work seemed like a good fit. Joan turned a spare bedroom into a home office, crowding the small room with a weathered brown chair, computer desk, and large monitor. Then she started searching the internet for work that she could do online.

Joan can't remember how she first found out about Amazon Mechanical Turk, but she suspects that she learned about it on Reddit. Reddit is one of several online communities where people doing ghost work share tips on how to get started. As a 39-year-old white woman with a master's degree in communications, Joan is, in some respects, a typical MTurk worker. Almost 70 percent of workers have completed a bachelor's degree or higher educational attainment. In other ways, she stands out. MTurk workers skew young: 76.9 percent are between the ages of 18 and 37, that bracket of years when people are most actively seeking their first career-defining job.

Though Joan doesn't recall all the details, the process for setting up a worker account hasn't changed since she first signed on. She would have gone online, navigated to the main MTurk website, and clicked on the sign-up button. As a newcomer, Joan would have been asked to enter a verifiable name, email address, and password. From that point, she would have been given access to the behind-the-counter side of the site. Visible from Joan's "dashboard" would have been dozens of tasks. Tasks, or what Amazon refers to as HITs (Human Intelligence Tasks), are jobs for hire. If she had clicked on a task, it would have shown her a short description of what the task required, the deadline, and what it paid. She could click on and complete a task, but, as a new user, she'd need to wait until her account was validated to get paid. Before Amazon pays a worker, it verifies the person's physical mailing address, national identity, and bank account information. That's how easy it was for Joan to join the ghost workforce.

To a new worker like Joan, MTurk's dashboard can look chaotic. One sees multiple expandable menu tabs, including a tab to keep track of one's account, another to track the individual tasks, and a tab that lists the worker's "qualifications." That word doesn't align with skills. In the world of MTurk, qualifications can be things like a worker's age, gender, and location. People who post jobs on Amazon use "qualifications" to restrict the type of worker who can accept the job. For instance, if an advertising company is looking for a focus group to give it feedback on a product meant to appeal to women in their forties, it might add qualifications such as gender and age to the job. It can even pay Amazon an extra premium fee for workers who list qualifications like "smoker" (30 cents) or "2016 voter" (10 cents). When Joan first looked at her MTurk dashboard, she remembers, she felt a bit overstimulated, but not deterred. "I thought, Okay, this is not going to pay out at the beginning, but if I do it for a while, it may become a decent source of side income," she says.

No one knows the exact number of people who use MTurk, but typically about 2,500 workers are actively either searching for tasks or completing tasks on the platform.[13] Because no agency—like a labor union or the Department of Labor—tracks this information, big-picture numbers are equally hard to pin down. Amazon maintains that it has 500,000 registered MTurk workers. According to researchers, anywhere between 100,000 and 200,000 people are registered to work on MTurk.[14] Panos Ipeirotis, a leading researcher most known for his work tracking the ebbs and flows of MTurk worker demographics, estimates that 2,000 to 5,000 workers can be found on the MTurk platform at any given moment. That is roughly the equivalent of a 10,000-to-25,000-person full-time workforce.[15] If we apply this logic to every on-demand platform, there are potentially millions of full-time jobs in the shadows of ghost work. This, of course, assumes that people would want to do this work full-time. However, as will become clear, a sizable percentage of workers stick with on-demand ghost work precisely because it does not demand a full-time commitment.

We posted a task on the MTurk platform to understand how workers are distributed around the world. Upon accepting the task, workers were shown a Bing map of the world and told, "Just double click your location and submit the HIT—it's that simple." Over ten weeks, 8,763 workers across the globe self-reported their locations. Workers are distributed throughout the United States in both highly and sparsely populated regions, but Indian workers are concentrated in the southern part of the country, a point we'll return to in the next chapter (see figures 1A and 1B).[16]

Like most workers we met, Joan starts her day looking for tasks. One of the tasks she does the most is text categorization. She might read a snippet of text, perhaps a sentence or two from a news story, and either create a category for it or pick "politics" or "sport" from a list of options presented to her. The first time we spoke with Joan, she was doing one such task. For every data point she categorized, she made two cents. She classifies tens of thousands of pieces of text every week.

Joan spent the first six months on MTurk finding her footing. In time, she learned that the trick to making decent money was to quickly find doable work and to evaluate the requester offering the job. She noticed that on MTurk, every second counted; a slow internet connection, time spent finding work, or any unplanned downtime was the equivalent of lost income. In her first year on MTurk, she made $4,400. Some people might see that number as insignificant, she says, but "$4,400 is a meaningful amount when your previous income was zero." Two years later, her MTurk earnings had almost quadrupled, to $16,000. Joan is now among the 4 percent of MTurk workers who are skilled, practiced, and lucky enough to earn more than $7.25 an hour completing tasks.[17]

Hypervigilance is a necessity for top earners. Those doing ghost work who make the most money spend hours monitoring their dashboards and scrolling through pages upon pages of job postings. Joan, like so many others who are trying to make MTurk a core source of income, turns to free software tools and workers' online forums to reduce some of the search costs that are an unpaid part of the job.[18] They must be ready to snap up a well-paying or fast-and-easy task the second it pops on their screen, lest another worker click on the link and accept it first. "I've worked harder at this than I ever did at any office job," she says. To enhance her speed, Joan arranged her web browser's display of the MTurk dashboard to show 25 tasks or HITs at a time, and she uses keyboard shortcuts she created to flip through the pages quickly.

When Joan is in the zone, she can complete about 1,100 HITs an hour, netting roughly $22 an hour. She knows that people are likely to assume that the work is mind-numbing, but she finds the variety of tasks intellectually stimulating. She especially enjoys work that involves editing, which plays to her strength as someone with a background in technical writing. "I'm good at it and it's easy to do," she says. When the work does feel mundane or repetitive, she stays alert by listening to techno music or watching television. When we spoke, she was working her way through several seasons of *Top Gear,* a show for car lovers. "People talk about 'Netflix and chill,'" she says, "but I watch Netflix and MTurk."

MTurk set a minimum fee for work at one cent per task and, from there, requesters decide how much to offer MTurk workers for each assignment. On average, requesters price tasks to offer the equivalent of $11 an hour, but low-paying requesters flood the market with minimum-fee work, which drags down the overall earning potential of workers, who must wade through lists of poor-paying tasks to find decent work. "It's a constant race to the bottom," says Joan. By some estimates, the total revenue of the requesters on MTurk and similar sites like CrowdFlower adds up to $120 million per year.[19] Workers keep what requesters pay out, but Amazon charges the requesters 20 percent of what MTurk calls "the reward" — a worker's paycheck, including any bonus amount (the equivalent of a tip) — as its fee for operating the platform. Amazon charges an additional 20 percent for HITs that require ten or more workers.[20]

Unlike a traditional employer-employee relationship, MTurk workers are largely anonymous and mostly autonomous, meaning that a requester cannot specify the people who will carry out the work nor dictate exactly how the task is completed once it's been accepted by a worker. Workers alone are responsible for the taxes on their MTurk income. They are expected to file as independent contractors, the 1099 forms familiar to anyone in the freelance consulting world. The trade-off for the requester is that the work is done fast and without the associated costs of officially hiring an employee. The trade-off for the worker is that they don't have to stick with the same job any longer than it takes to complete the task. They can fit work around the demands of their lives rather than hand their lives over to the long commutes or hostile environments that come with some nine-to-five jobs. And they can stop working the second they've made the money they need to make. But the completion of a task does not always equal a payday. The work submitted by MTurk workers is reviewed by a human being or an algorithm that either deems the work satisfactory or rejects it. If the work is rejected, the worker isn't paid. Each worker's approval rating, the fraction of tasks they have had accepted, serves as a reputation score on the site. Many tasks on MTurk require workers to have approval ratings of over 95 percent, so even one rejection can seriously affect a worker's ability to earn money by limiting their access to future tasks.

Like all people doing ghost work, Joan must weather fluctuating income streams. Requesters can bring booming business one day and disappear the next. Not long after she'd signed up with MTurk, Joan got a string of decent-paying tasks posted by Taste of the World, a pseudonym widely rumored among workers to belong to the popular travel site TripAdvisor. Taste of the World posted hundreds of thousands of tasks on MTurk, jobs like removing duplicate hotel listings, validating website links, writing descriptions of top travel destinations, creating city-specific lists of the best places to eat, and cleaning up typos. The average Taste of the World task could net an experienced worker the equivalent of $10 an hour, and working for the requester had other perks, too. "The work was available just about every day . . . and it was posted hours at a time," says Joan, meaning she didn't have to jump on it before it disappeared. She could step away from her computer to make dinner and, when she returned, the Taste of the World tasks were still available, because the sheer volume of work was so large. But just as abruptly as the jobs arrived, they dried up. Joan told us that less than a year into using MTurk, Taste of the World posted to MTurk Forum that "we have enough people." Joan flatly added, "And that was the end."

When it comes to paying for ghost work, each platform operates a bit differently. Amazon, in some ways, operates like both an ATM and a company store. New workers on MTurk must wait out the initial ten-day holding period before they can claim any money made doing tasks. After successfully submitting ten days of requester-approved work on the platform, U.S. workers have a choice: they can receive the full value of their earnings in the form of an Amazon.com gift card or they can transfer their paycheck into an Amazon Pay account. From an Amazon Pay account, workers can then transfer their earnings to a personal bank account, but they have to pay a transfer fee to Amazon for the privilege. International workers, with the exception of citizens of India, can only convert their earnings into an Amazon.com gift card.

Indian citizens are the only international workers who can also earn cash for their MTurk ghost work. There's no reason for this other than the fact that Amazon's multinational holdings allow it to operate and transfer money between its U.S. and India office locations. India's MTurk workers could opt to fill up an Amazon gift card, though the company doesn't reliably deliver to many of India's sprawling, informal neighborhoods. If a worker in India wants to move their money to a personal bank account, they must first hand over their birth date and a scanned copy of their permanent account number (PAN) card, the equivalent of a U.S. social security number. It takes Amazon a week or more to verify PAN card information. Once that is done, workers in India have one more bar to clear: they must send their bank account information to Amazon for verification. Once it's verified, Amazon can, for an additional fee, cut paper checks or offer direct deposits to India's MTurk workers.

For her part, Joan didn't plan to turn MTurk into a full-time job. It just happened. Now she has settled into the life of an independent worker, and her long-term goal is to create financial stability by knitting together several sources of income. This was a common theme among workers we met. And, indeed, 75 percent of workers on MTurk report having at least one other source of income. In Joan's case, in addition to her work on MTurk, she spins her own wool and sells knitted crafts at a local market. She's also ramping up her technical writing skills, with the goal of having a more competitive freelancing profile on the macro-task site Upwork. And she thinks about getting a part-time telecommuting job, like online customer service work, but she hasn't figured out how to do that while taking care of her mom. And, like 75 percent of MTurk workers, Joan does ghost work on other platforms, including Microsoft's UHRS, even though she notes that her primary source of income, for more than a year, was MTurk.

## UNIVERSAL HUMAN RELEVANCE SYSTEM: MICRO-TASK GHOST WORK BEHIND CORPORATE FIREWALLS

Before she had children, Kala, 43, whom you met in the introduction, worked as an electrical engineer. She stepped out of the workforce when her second child came along. But after staying home for several years, she missed the sense of community and purpose she'd enjoyed at work. She broached the subject of going back to work, even part-time, with her husband, but he was skeptical of her ability to juggle family and work. "He worried that it would be too much," she says. But Kala pressed. Finally, she and her husband landed on a compromise: she'd work from home. Now she works on Microsoft's proprietary platform, UHRS.

Microsoft, like several of the large tech companies, has its own internal micro-task ghost work platform, modeled on the mechanics of MTurk. Companies pushing technological innovation need legions of workers to beta-test products and check code. They also rely on people doing ghost work to improve their services' algorithms and artificial intelligence by cleaning up training data from large stores of proprietary data. Tech companies collect and archive information about how people use their sites. Data such as top search-query terms, popular song choices, and mouse cursor movements can be harvested to fuel product development. If customer data is the new oil, the people doing ghost work operate the rigs.

The biggest difference between MTurk and tech companies' internal platforms, like UHRS, is that MTurk recruits and sells labor as well as the platform work site itself, while, on big tech companies' platforms, a third party — a vendor management system (VMS) — recruits and supplies ghost work labor. All of this is to say that vendor management systems create yet another layer of opaqueness, acting as a broker finding people willing to do ghost work on contract, under nondisclosure agreements.[21] For example, Google used vendor management systems to populate its enigmatic ghost work platform, EWOQ. People hired through contract staffing companies like Leapforce worked on EWOQ, identifying and ranking new webpages to fine-tune links between ads and users' search queries.[22] According to press accounts and people we interviewed and surveyed who work on multiple on-demand platforms, Twitter and Facebook use internal tool kits that function much like MTurk and plug VMS-provided workers into their platforms to monitor and review content.[23] Kala does similar tasks on UHRS for Microsoft.

Before Kala started doing ghost work on UHRS, she worked for a small company that processed back office files from a U.S. business. Her company, one of many so-called business process outsourcing (BPO) shops handling work from the United States, was located in the heart of Bangalore's Electronic City neighborhood, not far from tourist-choked Cubbon Park.[24] Perhaps it is ironic that the company's biggest contract was with one of the oldest and largest labor organizations based in the United States. Kala and three other women sat shoulder to shoulder in the BPO shop's cramped four-desk office and did small tasks like deleting duplicates and fixing typos in data entries and updating the labor organization's contact database. She enjoyed going on a virtual scavenger hunt for information, like the correct spellings and zip codes of cities like Chattanooga and Hoboken that are scattered across the web. She liked tracking vital clues, entering the correct search terms, and finding information pertinent to the task at hand.

That's how, three years ago, Kala found her way to Microsoft's internal platform, UHRS. She started by searching Glassdoor, a job review site, where she found a link to a vendor management system. "I clicked that link. It seemed like it offered good ways to keep up my web search and other computer skills."

Kala qualified to work on UHRS with ease. Each vendor management system has its own procedures for vetting potential hires for ghost work, which often come down to language proficiency (the ability to write in English is often privileged) and prowess in finding materials online. On UHRS, for example, if an applicant passes a short quiz testing their language and web search skills, within a matter of minutes they're assigned a unique account. (For UHRS, a worker account is no different than a Microsoft account used for logging on to Xbox.) Once the account is active, workers can complete the platform-specific training, learning the ins and outs of UHRS's equivalent to the MTurk dashboard, and can start looking for projects, called "HitApps," on UHRS.

Which HitApps are available depends on the worker's current IP address and the country and language on file with their VMS. Once a worker on UHRS has completed at least one job in the HitApp (called a "Judgment") and shown they can do it correctly, the remaining HitApps are transferred from the Marketplace area into the worker's My HitApps area, the home screen where they can see micro-tasks available to them. Kala set up a desk in the corner of her bedroom and got to work.

The only people able to submit a request for ghost work on UHRS are Microsoft's full-time employees and authorized partners working with Microsoft to develop new products for the company. That means that Microsoft's more than 120,000 full-time employees can, at any time, become requesters in the loop commissioning workers in the ghost economy to help them with their tasks.

In the same way that Joan shared some qualities with her peers on MTurk, Kala has some similarities and some differences with her peers on UHRS. On UHRS, nearly 80 percent of workers are between the ages of 18 and 37, and more than 70 percent are male. But, like Kala, more than 85 percent of workers have a bachelor's degree or higher.

Much like the case of MTurk, there are no clear labor laws governing who can sign up and do paid work on UHRS. But ghost work for any multinational, like Microsoft, is by necessity a global enterprise. The availability of micro-tasks to workers on UHRS around the world largely revolves around Microsoft's immediate needs to support a range of products and services delivered to more than 20 countries in 70 languages.

The types of micro-tasks available to workers on UHRS are not surprising if you think about the products that Microsoft sells. Workers review voice recordings, rating the sound quality of the recorded clip. They check written text to ensure it's not peppered with adult content. Another popular task is translation. Microsoft's strength in speech recognition and machine translation comes from the ghost work of people training algorithms with accurate data sets. They create them by listening to short audio recordings of one sentence in one language, typically English, and entering the translation of the sentence in their mother tongue in an Excel file.

Other common types of work on UHRS are market surveys—often restricted by demographics like age, gender, and location—and a task called "sentiment analysis." In sentiment analysis, workers may look at a series of words, selfies, videos, or audio files and add a word to each data point that describes their sense of the mood of the word, person, action, or sound in front of them. These human insights become the training data for algorithms later shown the same materials.

Back at home, Kala often turns to her sons for help completing categorization tasks, especially ones that require knowledge of American colloquialisms. The boys help her categorize and sort the best terms for finding common websites. (For example, if a person wants to find an expensive wedding gift, would they enter "fine china" or "fancy dinnerware"?) Kala's children also help when she picks up tasks identifying "adult content," a common job that information studies scholar Sarah T. Roberts refers to as "commercial content moderation."[25]

This kind of content moderation requires someone like Kala in the loop precisely because words, as plain as they may seem, can mean many different things depending on who is reading and writing them. Artificial intelligence can learn and model some human deliberation like that between Kala and her sons, but it must be constantly updated to account for new slang or unexpected word use.

One of the points of friction when dealing with a vendor management system is that they are limited in the amount of technical support they can offer when someone has a question or something goes wrong. The most a worker can do when a problem or question about tasks comes up is click on the "Report a technical issue with this HitApp" link in their MyHitApps area, enter their issue in a text box that would be familiar to anyone who's typed in a service complaint online, then wait for someone to reply to their email with help.

Kala echoed what other UHRS workers talked about in the online forums: the MyHitApps area might be down at any time, with no explanation posted, locking workers out of their tasks. Because the VMS doesn't operate the platform itself, it can neither fix the glitch nor explain what's going on. And when this happens, UHRS engineers are always too busy getting the labor platform back up to respond to queries from people doing ghost work. Even though the engineers are sympathetic to the workers' plight, it's contractually not their job to respond to the worker side of the equation. Dealing with worker frustration becomes a giant game of "not it" between the VMS and the tech company using the VMS's ghost workforce.

Likewise, if there's a dispute about the quality of a worker's submitted tasks, VMS agencies don't intervene. Written into most ghost work contracts is that the workers bear the full responsibility of arbitrating disputes. Vendor management systems do typically provide forums to workers, though they are siloed by nondisclosure agreements. Workers signed up for tasks on Google through EWOQ cannot see or talk with workers picking up tasks for Microsoft through UHRS. But if workers have questions or problems specific to UHRS, they can turn to an active group of people doing ghost work talking about the jobs on UHRS.

Finally, new workers on UHRS must wait three weeks before their first paychecks arrive. Tech companies typically pay the VMS, which, in turn, pays the workers. Like MTurk, UHRS uses the first few weeks of a worker's presence on the site verifying their account information, checking their work, and setting up the money transfers that will go to the worker's VMS user account. After the initial waiting period, workers receive payment every two weeks.

Kala's in-laws disapprove of how much time she spends on the family's computer. They'd prefer if she spent more time with them, she says. But she enjoys the independence and having a little extra money in her pocket. For Kala, most work on UHRS means constantly learning new things. It also means staying connected to the IT sector and a particular job. Working on UHRS gives Kala confidence that she knows the latest software and strategies for finding information online.

Kala can also use her work on UHRS to fill the years on her résumé that might otherwise suggest that she left the workforce altogether. "It is hard for women my age to return to work or break into a new job. Everyone assumes that you only know how to chase after babies or you won't be able to keep up with what you need to know." And her husband has come to appreciate her work and finds ways to support her. She smiles recounting how he brings her tea and snacks in the evening when they are both home. "I love that he sees me working and does little things for me, like I do for him."

Although none of Kala's old office mates at the BPO followed her into ghost work, she does talk with them about how to improve their search query skills. The women trade tips once a week when Kala travels downtown to meet up with her former colleagues so she can feel, as she puts it, "part of the working world."

MTurk and UHRS illustrate how much micro-tasks still depend on human creativity. But the same approach to breaking down work can also be applied to larger projects. As the next two cases show, figuring out where and when to place humans in the loop opens up its own business opportunities. These companies blur the lines between micro-tasks and macro-tasks. In doing so, they also open up the alarming question: What job can't be turned into ghost work?

## LEADGENIUS: BLURRING THE LINES BETWEEN MICRO AND MACRO GHOST WORK

Zaffar, 26, has spent his life in the heart of Hyderabad's one-hundred-square-mile "Old City," a walled neighborhood along the banks of the Musi River. Built five centuries ago, it is one of the largest, oldest Muslim neighborhoods in all of India. His father helped build the freeway that rings the edges of the densely populated HITEC City, an IT industrial park built in the 1990s to accommodate the flood of outsourced work sent to India's shores. Because the first boom in IT work opportunities typically flowed to the upper castes of Hyderabad's Hindu majority, many of Zaffar's uncles and cousins joined part of a wave of Muslim Indian men emigrating to the United Arab Emirates, on the Persian Gulf, to find higher-paying work as drivers, line cooks, or salespeople in the shops crowding the emirates' beaches.

But Zaffar's father wanted a different life for his two sons. He pushed them to get college degrees and white-collar jobs. Zaffar's brother studied finance and works as a teller at a bank in a wealthy suburb of Hyderabad. Zaffar studied IT, getting the equivalent of a bachelor's in engineering from a local technical college.

Tech company jobs in Hyderabad go to candidates with polished English, especially if they've mastered British or U.S. English intonations. Young Muslim men who have few chances to practice their spoken English skills and who have limited engineering training, like Zaffar, are at a disadvantage. He spent about a year applying to call center positions and tech support positions at several of the large tech companies based in Hyderabad but never made it past a second interview. So when he saw a news article about the on-demand platform LeadGenius, he applied.

LeadGenius is a business-to-business service selling leads to salespeople. Other companies sell sales leads, but the genius in LeadGenius is the creativity and insights of its workers. Think about it this way: a basic web search can return contact information for potential new customers and clients, but artificial intelligence can't determine whether the information is useful. That's where people come in. A person can look at the information about any two businesses—for example, how long they've been open, whether they have other store locations, or if either business owner is in the middle of a lawsuit—to help a salesperson decide which of those businesses is more or less likely to be a good target for sales. By cultivating a workforce that feels seen and valued in ways that are hard to come by in on-demand work, LeadGenius bridged the chasm between micro-tasks (small, repetitive, and somewhat mindless work) and macro-tasks (work that requires thoughtful sleuthing).[26]

Zaffar found that signing up with LeadGenius worked much like it did on other ghost work sites. As a newcomer, he would have clicked on a button that said "Apply" and created a "candidate account." He would have taken a typing and proofreading quiz and answered a few demographic questions — age, gender, location. He would have submitted a résumé. LeadGenius makes a point of letting applicants know that informal work experience, like working for one's family, is legitimate — a small gesture of recognition that many of its recruits live in places where work is less than formal.

Where LeadGenius's hiring practices diverge from those of other on-demand platforms is what comes next — a rigorous interview round conducted by the workers who've advanced up the ladder. The entire process can take up to three weeks. A job invitation is offered only when an applicant passes the interview, as well as some additional tests.

If the job offer is accepted, the new hire takes part in a paid video orientation and post-video quiz. Paying workers for an orientation also sets LeadGenius apart from many of the other vendors and open ghost work platforms. All new hires start out on a 90-day trial, but if they make it through the first 90 days, keep up their requirements by logging in and staying connected to teams for at least 20 hours a week, and make it to their shifts on time, they get an automatic bump of 8 percent in their hourly pay.

In exchange for a 20-hour commitment, workers must provide their own computer and internet connection, be able to work with office software like Microsoft Word, Excel, and Google Docs, and be comfortable using instant-message and voice-chat software, like Skype. Zaffar works on a laptop he bought himself. He likes to move to different parts of his house, to break up his work shift, rather than sit anchored to the desktop computer that he set up in the foyer that serves as his home office.

Like Zaffar, 85 percent of workers on LeadGenius are between the ages of 18 and 37.[27] Slightly more than 70 percent of LeadGenius's ghost workforce — called researchers — have at least a bachelor's degree. Globally, women make up 49 percent of the platform's workforce, although among our surveyed India workers there were 10 percent more men than women. Almost 75 percent of workers on the platform use LeadGenius and at least one other platform to do on-demand work. When Zaffar first applied to LeadGenius, he was still working full-time on MTurk but hadn't made his daily goal of $20 in more than a month. According to LeadGenius, one out of every three researchers supports a household of three or more people. And we found that more than 60 percent of workers on LeadGenius rely on the platform, in addition to at least one other income stream, to meet their basic needs.

LeadGenius has a global workforce. The company's largest research teams are based in India and the Philippines. Wages depend on the business markets for the sales leads that LeadGenius is hired to find. As with UHRS, then, workers' pay depends on a mix of two things: first, what's called labor arbitrage — how cheaply a business can get work of comparable quality out of workers connected to global trade around the world but living in a country where the wages are lower — and, just as important, the value of "localization" to a business's product or service. As more and more companies try to sell what they make to a global market, they create a demand for workers who know the local language, idioms, and quirks of their corner of the globe.[28]

LeadGenius structures its research teams like a traditional workforce. Positions with stair-stepped levels of responsibility — trainer, junior manager, and project manager — create a company-wide trajectory for career growth. Workers at higher levels assemble teams to tackle specific client requests, which often looks like waves of web searches to collect, sort, and refine sales leads.

Teams are organized so that all the members are working in the same time zone, typically in the same country. Project managers also reside in the same time zone, so they can answer questions. Workers must be able to devote between 20 and 40 hours a week. Once they join a project team, they are asked to stay with it for a minimum of 30 days, unless an emergency arises. Workers are evaluated by project managers. There's room for a learning curve, within reason. They receive a strike if they return work that's incomplete or done incorrectly. After three strikes in three months, workers are removed from the platform.

Workers have a dashboard for most projects, where the lead requests are preloaded, just as they would be for MTurk or UHRS. In some cases, LeadGenius's business clients have workers come to their internal websites for privacy and security of the data. The dashboards are also the hub for all communication among workers, managers, and those clients connecting directly with the projects. If questions pop up, workers can turn to a manager for help, and workers have live-chat software running in the background, enabling them to talk to one another, just as they might if they were working in a retail store. Zaffar spends hour after hour making tough, quick decisions about which leads to send to his team.

Since it's a business-to-business company, the daily work can be hard to picture, so here is one example. Law offices in the United States pay platforms like LeadGenius to collect the names of people found in public records. Workers like Zaffar take a city, like Cambridge, Massachusetts, then go online and search local newspapers for postings of people who broke the law. Maybe they were arrested for driving under the influence or maybe they defaulted on their alimony. Those are the kinds of sordid details that may show up when a potential employer looks up a job applicant online. Workers for LeadGenuis compile an exhaustive list of people who've been publicly listed for allegedly breaking the law, and then they hand over this list to a lawyer, who will begin calling the people on it, offering to expunge their record from search engine results in exchange for a fee. As one worker told us, this is the kind of lead generation that is constantly providing new work, because "people will always commit crimes that they want to cover up later." You might say this is a new form of ambulance chasing that comes with the internet. It requires smart searching techniques that a computer can't do on its own.

A worker's day ends when the time they've listed for their availability runs down — what LeadGenius labels as a shift. And, much like we found among other workers, LeadGenius's team members end their shifts and meet up with fellow workers who live in their local communities.

LeadGenius pays workers every other Tuesday, by noon California time (that's where the startup's headquarters are based), using digital payment services PayPal, Payoneer, and even Bitcoin. But, as with MTurk and UHRS, getting paid requires direct deposit, so workers must take a leap of faith and connect their bank account to accept a global transfer from a company headquartered on the other side of the planet, operated by people they may never meet or speak to in person.

Working for LeadGenius, Zaffar saved enough money for his marriage ceremony and to take almost a month off from work. He also was able to step out for more than three weeks to nurse his mother back to good health after a debilitating auto rickshaw accident, and, just as quickly, he returned to LeadGenius. They had guaranteed him a spot if he wanted to come back. Zaffar turned down a junior manager position with LeadGenius right before he got married, because his fiancée did not want them to start out married life with Zaffar working the night shifts or 30 hours of work required of the position.

## AMARA: TRANSLATING LANGUAGE INTO GHOST WORK

Karen, 37, has a bachelor's in comparative literature. She lives in Portland, Oregon, with her husband, three-year-old son, and ten-month-old daughter. Amara, a web-based interface for linking video to captions, wasn't Karen's first online job. In years prior, she'd taken on work through Lionbridge, a vendor management system, doing search engine evaluation for (she suspects) Google. She also worked for Fancy Hands, an on-demand virtual assistant service. The work involved fielding live webchat and text-based requests from people asking for help with anything from scheduling a flight to purchasing furniture. But both of those jobs, which demanded constant searching for tasks, were short-lived.

After her second child arrived, Karen started looking for work that was more creative. She began writing and editing "how to" articles for a media company (what she called a "content farm") with clients like eHow and Livestrong.com. Karen says she was "fired" from the media company after several tense email exchanges with supervisors who marked her work as "unsatisfactory" but didn't offer feedback on what she should do differently with the next assignment. One cannot, technically, be "fired" as a freelance worker, but, as Karen says, "no matter its legal status, it felt like getting fired." Then, searching Craigslist for other copyediting jobs, she came across an ad for Amara.

Amara is a translation and video-captioning service that blends the mechanisms of ghost work with automated features that manage the slicing and recombination of translated video content. Amara blurs the boundaries between repetitive micro-tasks and larger macro-tasks that call on a worker to bring creative insights to a project. The work available through Amara also challenges the clear lines that most in society would draw between paid work and volunteerism, or a labor of love.

The idea for Amara dates back to 2006. It was the brainchild of friends Nicholas Reville, Tiffiniy Cheng, Holmes Wilson, and Dean Jansen, all then working with the nonprofit Participatory Culture Foundation (PCF).[29] The group had a small amount of grant money to build tools that helped people more easily share videos and creative work on the internet with no gatekeepers or advertisers controlling content. At the time, Real Player and Windows Media Player were the only choices for distributing video online. Building on its early software tools, PCF launched Amara in 2011, a web-based platform that helps people collectively add language translations to dialogue and scenes in videos playing on their screens.

In Spring 2011, not long after PCF released Amara online, activists turned to it to translate videos documenting human rights crises, most notably during the Arab Spring and the Fukushima reactor meltdown. This launched Amara into the limelight. Filmmakers and the nonprofit Technology, Entertainment, and Design, the creators behind TED Talks, approached PCF looking for ways to offer "rush captioning" to media creators and TED presenters who want to caption video for a global audience.[30] By mid-2013, PCF Executive Director Nicholas Reville and seasoned technology strategist Aleli Alcala co-founded Amara On Demand (AOD) to fill this niche. Amara represents two realities folded into the growth of ghost work. First, ghost work's simple existence and persistence belies claims that it's ever easy to completely jettison humans from workflows that require creativity. Second, Amara speaks to a nascent desire among some businesses to explicitly acknowledge that people, rather than software, are the more valuable component behind ghost work.

Karen started out on Amara as a volunteer, subtitling YouTube videos, short documentaries, and college lectures for deaf and hard-of-hearing communities. The macro-tasks appealed to her for reasons that we heard from other Amara workers as well. She enjoyed being a stay-at-home mother but was eager to find avenues for adult interaction with her co-workers. When she moved to working with Amara On Demand, the work was already familiar, but now she was getting paid.

Amara's pay rates vary according to the demand for a specific language. Captioning and translating content in languages that are more commonly spoken, particularly those that map to a wealthier country, fetches a higher premium. So, for example, Amara can pay Karen around $68 per hour of video to caption what she sees and hears into written English. The first time Karen tried to caption a video, it took her an hour to complete a mere minute of material. But she practiced and got faster, and the captions started to flow more quickly, bringing her pay up to the average.

To that end, Amara's workers captioning English content make about a dollar for every minute of video they caption. Now Karen works on about 15 minutes of video at a time, but that earns her a little over $15 an hour, almost twice as much as she would earn making coffee drinks at her local Starbucks. And Karen doesn't have to commit to a full-time job to get that pay rate. But the best part, according to Karen, is the teamwork.

In her earlier forays into online work, Karen worked alone. Now she collaborates on video-captioning projects with a team. Amara On Demand organizes teams working on a single video into small groups, the size being dependent on the language and size of the translation project. If it is a feature-length film heading to an international film festival, a job Karen recently took on, Amara will assign a few team leaders to supervise the project, send out job requests, and assemble the team.

Once in place, everyone receives an official invitation to work on the team. Then interested team members send back the times they're available to work on the video assignment. Once a team member accepts the invitation, they can choose any of the videos available and share clips or notes with other people on the team. Amara's on-demand teams work mostly as equals, both producing original subtitles and editing the translations that other team members produce.

Nearly 75 percent of the people working on Amara are between the ages of 18 and 37. More than 60 percent of Amara's on-demand team are women (the reverse of all the other platforms that we studied).[31] More than 78 percent of Amara members hold a bachelor's degree or higher (and more than 40 percent working on Amara hold a master's degree or higher). Eighty percent of Amara workers rely on the platform, alongside at least one other source of income, to meet their needs. And for nearly 70 percent of workers, Amara is the only ghost work platform that they use to access on-demand work.

Amara still operates most of its team communication via email or a live-chat channel. Team members can create profiles so that their name, picture, and bio can be shared with those with whom they're working on a project. The overall tone among team members is amicable; some people even go so far as to email friendly notes to each other, she says. All and all, it's a very different vibe from her earlier experience working on demand.

But there is another big difference between working for Amara On Demand and working for other platforms. Amara makes it easy for team members to return a task (a video) if they start it and then realize it's either too demanding or not interesting to them. Karen recalls rejecting a task only once after accepting it. It was a video recording of a Samuel Beckett play. "Amara needed subtitles for it. The work was so daunting! The characters talked really fast, and the dialogue was, literally, 'Blah, blah, blah, blah, blah.' Somebody had to caption it, but I mean, I can't imagine." Karen was paid for the amount of time she spent trying to figure out how to do the subtitles: exactly five minutes of reviewing the video. The structure and ease of picking up as well as dropping projects encourages team members to experiment with topics that might otherwise feel too intimidating.

Amara took pains to build software that would not only be easy for workers to use, but fun. After all, it needed to first appeal to volunteers. Amara On Demand team members are supplied with all the software they need to do the captioning and translation of videos. Jansen and Wilson used the popular dance video game *Dance Dance Revolution* as a model for adding captions to videos and adding features like big, colorful buttons and a simple interface. The software makes it easy to use the tab key to browse the video materials. As a team member watches a video clip, they type translations of the dialogue or descriptions of the action into a text window on the screen. They can then click and drag their subtitles into the Amara Editor, attaching their captions to that video segment. They can also stop and start a video as often as they need. "You're just basically hitting two keys to either start or stop your subtitle to sync it up with the video. It's a great program." You could say that it's as easy as playing a video game.

Automatically recognizing and translating language looks easy in some ways because people are accustomed to the everyday nature of tools like Siri, Cortana, and Alexa. Automating human speech recognition and translation is a fundamental part of artificial intelligence that grew into a field called natural language processing. Natural language processing was helped immensely by the internet's capacity to amass tons of examples of people writing and speaking in various languages. Yet capturing dialogue in video, particularly action scenes that change the mood and meaning of an actor's words, remains a difficult task for a computer program to understand, let alone translate into different languages. In fairness to the computers, it takes a team of people to achieve this, too.

Amara, as a nonprofit, challenges the unbridled venture capital enthusiasm for companies like Uber and the wave of startups in their wake that have pitched themselves as part of the "Uberization" of their market. Together, LeadGenius and Amara represent business models willing to concede that they don't just sell matching software. They are in the business of banking on people's creativity. They also use ghost work to deliver on tasks much larger than tagging an image. These macro-tasks are, at least today, out of automation's reach.

## UPLOADING WORK: WHEN FULL-TIME EMPLOYEES MANAGE MACRO-TASKS

In matching workers to another company's tasks, LeadGenius's and Amara's business practices can't help but draw attention to the other humans in the loop here, who are also hard to see. They are the requesters — small sole proprietors of small businesses or full-time employees at larger companies — managing their own workloads by finding and hiring help on ghost work platforms. They are yet another set of humans entangled in ghost work's loop, working to clear time-sensitive or new projects from their desks.

Commercial platforms like Upwork and a growing crowd of competitors take a hybrid approach to managing ghost work. They allow API access, which enables automated hiring, evaluation, and payment, not unlike MTurk. But they also allow individuals or companies to place tasks on the site manually and interact more with those contracting workers to complete a chore that looks more like something that might happen a few cubicles down than highly atomized micro-tasks. Common tasks placed on such sites include graphic design, video production, and content creation but also more engineering-focused tasks like website production and software engineering, some paying $100 per hour or more. Larger, more complex tasks require more interaction, so Upwork allows requesters to chat in real time with and send email to workers, which is useful for more complex tasks. Nonetheless, the platform still brokers the interaction between the worker and the requester. In doing so, it also distances the two parties. This can dehumanize the worker in the requester's eyes, as though the worker is simply part of the platform's software. Platforms like Upwork show that automated and semi-automated processes now distribute and manage the full range of work, from simple to complex.

Perhaps it's ironic that we met people hiring workers on sites like Upwork who turned out to be full-time employees themselves. They turned to ghost work platforms to subcontract out a variety of macro-tasks driven, in most cases, by four reasons familiar to anyone who's ever felt understaffed or overwhelmed at work. First, just as they might post a temp job ad on Craigslist or Monster.com, full-time employees used on-demand platforms because their own firms did not have the expertise that they needed for a project in-house. This expertise ran the gamut, from writing copy to analyzing vibrations in engines. As one senior manager at an engineering firm explained, he typically hires on-demand workers to do detailed engineering designs. "I was looking for a vibration specialist on a large-size induction motor. This is where I was looking for someone specializing in designing as well as engineering." A marketing manager at an online education company who uses on-demand workers primarily for content design, writing, and animation said, "If we can't make it work in-house, we look for a freelancer to complete the job. But there are also instances where we just don't have the skills, so we have no choice but to hire a freelancer."

Second, hiring on-demand workers can be done much faster and with lower costs and lower overhead than hiring via a traditional staffing agency, which is the most attractive feature for companies trying to maximize their profit and increase their bottom line. A communications specialist at a marketing company specifically called out and estimated the savings in hiring on-demand workers: "The most significant aspect of working with a freelancer is that they deliver value at a minimal cost. We can save up to 40 percent by not paying benefits or allocating office space." The marketing manager could also estimate the savings right off the top of their head: "If you go to [a staffing] agency regarding a project, they will charge, say, $2,500, but if you go to Upwork, you can finish that project for, say, $700 to $800."

Third, one of the most common reasons cited for using on-demand work is an unexpected spike in workload. In these scenarios, full-time employees of the company may be busy with other work when a new task arrives, so the firm hires an on-demand worker as an extra pair of hands. The marketing manager said, "We hire freelancers when we have a quick turnaround project. If our in-house team isn't available or we're getting slammed with a lot of work at once, we bring freelancers on board." Full-time employees hiring out ghost work reported a variety of reasons behind the increased workload, including seasonality or a rapidly approaching deadline, and in some cases both. A project manager at a direct mail company said, "We're a cyclical business, and even seasonal, so during peak times everyone is working at full capacity. That's when we raise our hand and say we need to hire a freelancer."

The final reason our interviewees said they hire on-demand workers is that they produce higher-quality work than contractors hired from a staffing agency or, in some cases, even full-time employees. A marketing manager at an advertising company said, "I found in a lot of cases that their performance was above an in-house employee." There are a few reasons for the higher-quality work. First, on-demand workers work for themselves and want to be called back for repeat work. We will see that a common technique among hiring managers for overcoming some of the shortcomings of hiring on-demand work is to maintain a trusted pool of on-demand workers they can repeatedly draw from. On-demand workers realize this, so they do high-quality work to make it into these trusted pools. It's as if every task from a new requester is a type of tryout to make it onto their callback list. A second reason for the higher-quality work output is the competition for jobs in the on-demand labor market. Jobs will go to the workers with the most up-to-date skills, whereas full-time employees can let their skills go stale without serious immediate consequences. A project manager at a healthcare company said, "I think some of the finest technical people come from the area of [freelancing]. It takes a special kind of person, who is diverse, who is more into learning more systems, new processes. I think [freelancing] sharpens your skill set, it keeps you sharp and on your toes . . . Their skill set makes them more marketable to a company."

Despite the sincere appreciation for workers' sharp skills and work ethic, there was also a tacit recognition among requesters, sometimes voiced, that they were outsourcing tasks that, under other circumstances, they might have done themselves. In plenty of cases, the distinctions among those generating macro-tasks and those completing them seemed categorically arbitrary. If there wasn't much difference between what a full-time employee did and what they might hand off to a worker, why did the full-time employee seem to have all of the perks and none of the risks of carrying out that assignment?

## Weaponized Ignorance

As ghost work swiftly and stealthily displaces a full-time workload, it is upending a century of efforts to keep full-time, long-term employment as the cultural default and stabilizing cornerstone of middle-class life.

Humans in the loop appear interchangeable. As noted earlier, thanks to application programming interfaces (APIs), workers are represented as a string of letters and numbers instead of a name and a face. In this dehumanized zone, few companies that sell ghost work have any idea who makes up their workforce ranks. Some of this erasure can be chalked up to logistics. The crowd is too big to see individual faces, one might say. But it's important not to gloss over the fact that erasure is a purposeful feature rather than an errant bug of the ghost work economy.

There are legal reasons why an on-demand platform might not want to know or care too much about its workers. But as the range of ghost work above suggests, it is hard to ignore how necessary humans have become in the growing shadow of AI.

CrowdFlower, a crowdsourcing and data-mining business founded in

2007 by Lukas Biewald and Chris Van Pelt, is the company behind Uber's "selfie security," discussed in the book's introduction. The company has other large business clients, like eBay, Mozilla, Twitter, and Facebook. People doing ghost work on CrowdFlower complete micro-tasks like approving photos, customer support, and content moderation. In 2012, a CrowdFlower worker, Christopher Otey, filed suit against the company, calling out its labor practices. Otey and a second worker, Mary Greth, were named as the suit's plaintiffs, but before it was over, an estimated 19,992 CrowdFlower workers had signed on to the lawsuit. The basis of Otey's lawsuit was that, according to Otey, CrowdFlower set expectations accordant with a full-time employee, but the pay and benefits were commensurate with those of an independent contractor. Otey says of his time working for CrowdFlower, "I didn't have control over the work I did. It was all done on their platform. I couldn't choose my own hours. I had to work when they provided the work. They pretty much controlled all the aspects of the work that was being offered." Given the degree to which CrowdFlower set the terms of employment, argued Otey, CrowdFlower owed him and other workers minimum wage, per the Fair Labor Standards Act. The company's legal team countered that, because CrowdFlower's workers were "free contractors," the FLSA didn't apply. Ultimately, in 2015, CrowdFlower paid $585,507 to settle the lawsuit, which left the question of the employment status of its workers unanswered.

Since 2015, companies buying and selling on-demand work have tiptoed cautiously around any activity that makes them look as if they are doing anything other than providing an online meeting place and matching service between people with jobs to be filled and workers willing and able to work. On-demand ghost work platforms see themselves as neutral parties, arguing that they are the software serving as the middlemen managing what economists call a two-sided market.

They connect requesters seeking workers, on one side of the platform's marketplace, and workers seeking jobs, on the other side. And in the absence of set hours, work sites, or agreement about who's the official boss in charge, it is difficult to gauge how much ghost work is done across this burgeoning industry, who's paying for it, and which workers are completing the tasks. The transaction costs that economist and Nobel laureate Ronald Coase so long ago identified as the reason for the very existence of firms seemed to melt away with the new on-demand systems. Platforms could keep themselves and requesters at arm's length from workers, shielded from a formal employer's legal responsibilities.

## WHEN YOUR WORK HAS NO CATEGORY

The paradox of automation's last mile suggests that the shift to using ghost work to deliver services is just heating up.[32] As of today, there are hundreds of companies offering on-demand ghost work services to evaluate, sort, annotate, and refine the terabytes of "big data" that consumers produce every moment they spend online, and an explosion of companies hosting larger tasks that are, at least in part, managed by APIs.[33] Still, treating ghost work as a consumable good drains ghost work jobs of any protections.

It can be difficult for those involved to fully see and value the expansive range of ghost work. Complicating the issue is that workers themselves don't know how to categorize their work or their status as workers, making it harder to figure out what people performing ghost work might want or need. Without a shared workplace, hours, or professional identity to orient them, those doing ghost work form casual, informal communities and social circles made up of diverse interests. This is a common feature of online environments where individuals drift in and out of their networks, drawn to different people and projects, depending on the time they have in their day. Organizing formal employment to equally prioritize and support each worker's ability to choose when they work, who they work with, and what projects they take on is unprecedented. Usually that privilege is reserved for the most elite full-time worker. Everyone else must cram their lives into the constraints of a nine-to-five grind or step aside for someone who will.

Businesses using a seemingly expendable, fungible contingent labor pool to knock out miscellaneous tasks outside the scope or purview of a full-time employee is not new. Ghost work is arguably just the most recent iteration of a well-established historical trend.

# 2

# From Piecework to Outsourcing: *A Brief History of Automation's Last Mile*

The National Labor Relations Act of 1935, authored by New York senator Robert F. Wagner, became the first piece of federal legislation in the United States to guarantee the basic right of workers to form unions, collectively bargain, and strike for better work conditions. Most think of the Wagner Act as the country's first publicly mandated social contract and safety net for modern employment. But defining workers' rights and weaving them into the nation's fabric actually started a century before that.

A thin, almost invisible thread connects the fates of 19th-century factory workers addressed by the Wagner Act and today's on-demand workers left unprotected by it. The ghost work chronicled in this book shares a legacy with New England farm families turning strips of cloth into fancy dress bows in the 1900s. They, too, more than a century ago, were largely left out of the Wagner Act's protections. This fault line—which divides full-time employment from everything else—also connects to young women in California calculating the optimal weight of jet fuel for rockets propelling satellites into space in the 1960s. And today the fault line stretches to the temporary labor pools around the world that, since the invention of the internet, have been tasked with managing databases and call centers and handling the accounting for *Fortune* 500 multinationals.

To understand why full-time employment remains the cultural measuring stick of success and how that weighs heavily on those performing ghost work today, we need to examine the past. This chapter returns to the end of the 19th century and moves across the 20th. Each moment of technological innovation that is highlighted shows how political leaders, economic power brokers, labor advocates, and the social norms of the day reproduced divisions between skilled professional work (meaning what is beyond the capacity of machines) and unskilled work (meaning contingent labor headed for automation).

The early days of industrial capitalism and manufacturing depended on assembly line workers to keep the factory running. But manufacturers also required a large number of people doing pieces of work by hand, tasks that could not be folded into the mechanical processes.

It took organized labor and the collective action of workers to make full-time employment in the semi-automated world of industrial manufacturing inhabitable. Unfortunately, the valorization and validation of full-time employment also made it easier for corporate interests to position piecework and, later, other forms of temporary or contract labor as expendable, that is, work that did not warrant protections. The investment in some workers and not in others hinged on who was doing the work and whether those jobs registered as "skillful" or seemed ripe for automation. So Cold War–era engineers, predominantly men and professionals with advanced training, were presumed to be inherently a more skilled and valuable workforce than the women "computers" filling companies' secretarial pools.

Companies driven, from the 1970s onward, by quarterly profit reports became more likely to see full-time employees as liabilities rather than assets.[1] Corporations used internet technologies, like distributed databases, and temporary staffing to outsource anything possible, and left full-time employees to wonder whether their jobs might go to globally outsourced contractors or "permatemp" staff sitting in cubicles down the hall. Corporate moves to greater reliance on contingent workers won out and redefined what counted as expendable labor in the process.

Workers' stories spanning the past century speak to what "full-time employment" means and where that meaning comes from. They also help explain why, when industries fixate on automating jobs away, they paradoxically spike demand for ghost work, shredding the social contract between employer and worker in their wake.

## BEFORE WAGNER: EXPENDABLE LABOR FUELING EARLY CAPITALISM

The Wagner Act was not the first employment law on the books in the United States. Prior to the Civil War, legal statutes codified who could be forced to work without pay through slavery and who had a right to be paid for their labor. Northern cities processing iron ore, cotton, and other raw materials grew wealthy off manufacturing goods subsidized by slave labor. Slaves, valued as property, were a fungible labor pool, expendable once they had served their work purpose. Capitalist expansion and Industrial Revolution growth, across every region of the United States, slave state or free, depended on the designation of slaves as a contingent workforce.[2] Only white male property owners could safely presume the right to demand compensation for their time.[3] Until the end of the Civil War, in 1865, that small fact—that a person commissioned to work had a right to collect wages—could not be taken for granted in the United States.

After the Civil War, even as industrial capitalism and opportunities to earn a wage expanded at a rapid pace, the vast majority of the war-torn nation lived in a twilight between subsistence agriculture and industrialization. One's own labor literally put food on the table. People eked out a living on small plots of land. Families sold what they could spare of their own yields, adding income earned from eggs, wool, or their expertise in sewing or in clearing land to make up the bulk of day-to-day commerce. Weathering droughts, deluges, poor yields, and famine was most of the population's full-time job.

The prospect of wage work prompted families to move from precarious subsistence farms, particularly in the South, to the flourishing port cities dotting the Eastern Seaboard. These cities represented the possibility of incomes that could purchase the comforts of lighting, heat, and food. New pools of cheap labor — free blacks from the South, and an influx of European immigrants — made it easier for northern factory owners to expand while keeping wages for workers low.[4] Against this backdrop of an influx of poor yet able-bodied groups of immigrants and former slaves and the young nation's first oligarchs funding modern industrial capitalism, an amalgam of assembly lines and piecework was born.

## THE INDUSTRIAL REVOLUTION'S TWIN ENGINES: ASSEMBLY LINES AND PIECEWORK

Placing workers along assembly lines did not outstrip the pace of organized trades and artisans overnight.[5] It took more than 100 years for factories with conveyor belts and mechanical pulleys snaking their way around workers standing in place and executing a discrete task to become the norm. Less often noticed is that, even after most manufacturing moved to mass production lines, some jobs remained, literally and figuratively, outside the assembly line process.

The first so-called manufactories of the Industrial Revolution could produce guns, locks, chairs, candies, shoes, and clothes at a pace 10 to 20 times faster than any guild or group of trade specialists working by hand. But in each case, these mass-manufactured versions of handmade consumer goods still required a person to put on the finishing touches. This was not a new facet of production. Guilds had long used apprentices and systems of subcontracted helpers to stretch the work of master craftsmen as far as possible. The new assembly line systems carried forward the practice of subcontracting that was ubiquitous among trade guilds, but with one difference: factory owners took control of both the raw materials and the labor "supply chain." Individual workers no longer coordinated production with one another. They were told where to stand, what to do, and how to hand off their contributions to be incorporated into the final manufactured product. In this way, the assembly line did not introduce a new division of labor between humans and machines so much as it wrested away control over the pace at which a person worked and their ability to coordinate with or delegate to others.[6]

As the Industrial Revolution got under way and machinery began to automate the production of certain goods, such as textiles, the availability of piecework exploded. Piecework (also called "industrial homework," "putting-out work," "cottage industry systems," or "commission systems") was the part of manufacturing or processing a product done by a person when a machine hit its limits. Piecework jobs were broken down into small, distributable tasks that could be carried out off-site, without stopping production or diverting resources away from the factory floor.[7] Assembly lines depended on old divisions of labor. Women and children living at the margins of cities dominated the expendable piecework labor pools.[8] In effect, industrial piecework was the first iteration of paid, on-demand ghost work.

Women and girls operated most of the ironing machinery and steam-powered sewing machines of the day.[9] They were the "finishers" sewing on buttons, curtain loops, shirt flourishes, and belts.[10] Finishing moved inside the factory's walls for only the larger-scale operations. Smaller-scale garment manufacturing, producing most textiles, remained dependent on farming communities taking in work through the "commission system."

The workroom wasn't a crowded tenement, the more typical scene of a "sweating system," as it was called then, or the sweatshops that still operate today. Instead it was a farmhouse with sewing machines taking up every inch of the living room, each family member, from fathers to small children, doing something with fabrics. Families paid a driver to haul sheets of cut cloth, giving him a percentage for every dozen sheets, based on the distance traveled. The driver became an intermediary for piecework, navigating the practically impassable country roads just within the city limits of industrial centers like Pittsburgh. They dodged the "wooden shacks swarming with chickens and children."[11] Drivers could hear the metallic clatter and whir of foot-powered sewing machines as they drove in from Pittsburgh.

Economists and industrial theorists of the day thought that piecework was a technological inefficiency that would fade out as mass manufacturing scaled up. For example, the spinning jenny, invented in 1770, was a single machine, powered by water, with as many as 120 spools hooked to a wooden frame cranking away to spin, draw out, and twist fibers. The machine could replace dozens of human hands working hundreds of hours to create the same amount of cloth for weaving. And Eli Whitney's cotton gin, invented in 1792, could deseed cotton and prepare it for the spinning jenny 25 times faster than a single person working by hand. Together, these two inventions ramped up production and mainstreamed consumption of cotton in the early stages of the Industrial Revolution. Most analysts tracking the growth of industrial production assumed that, between mechanization and the scientific application of technologies to manage a more skilled workforce, pieceworkers would eventually disappear.

None of these analysts considered how automation might create a short-term boom in the demand for contingent labor. Take the case of cotton. U.S. slave owners needed five times the number of slaves by the start of the Civil War because of the spike in demand for cotton and the desperate need for humans to continue to do what machines could not. Technologies like spinning jennys didn't eliminate the need for human labor so much as they repurposed and shuffled labor demands to a new set of temporary workers. Children became valuable pieceworkers in textile mills, because their small hands could reach between moving spools to clear lint and other debris that slowed down the machines. But the capacity to work alongside these machines, bridging the gap of automation's last mile, was written off as "unskilled."

Working the spinning jennys was considered manual labor that required no thinking at all, even though early accounts of children deftly moving from one heavily vibrating machine to the next indicate that the work took both mental and physical finesse. Surely piecework would evaporate as factories mechanized, and what was once guild-driven artisanal craft production and child labor would become a thing of the past. Or, as was happening all over Europe, unions would eventually bring pieceworkers into the fold by blocking workers from subcontracting out their own labor, just as they had under the skilled-crafts-guild system. Yet neither challenge to piecework ever dislodged it fully from the production cycle, particularly in smaller manufacturing settings that couldn't afford the newest machinery.[12] As one Pittsburgh garment factory owner put it in 1907, "It would cost us far more to have the women inside. We would have to rent another floor to make room for their machines; we would have to buy the machines, and we would have to have gas and heat, and then have to pay them more, may be, into the bargain."[13]

Unions proved themselves no more motivated than the factory owners to recognize the value of women's contingent labor, either on-site or working from home. The largest, most progressive union, the United Garment Workers, organized in 1891, tried to root out what it saw as the "menace of the outworkers" and make them "a coherent part of its growth," but to no avail.[14] Union organizers focused on getting young women to fill vacancies on the factory floor. But these approaches did not contend with or even recognize how often women worked on contract through piecework, because factory work was still considered morally suspect for a young, unmarried woman and, practically speaking, took her away from her other full-time job of cooking, cleaning, and caring for children and elders at home.

Young women could earn twice their week's earnings of $4.50 and leave the backbreaking work of sewing machines behind if they worked at a UGW union shop. But union strategies did not prioritize or recognize the specific burdens or costs women faced leaving contract labor or home-based piecework behind. Unions quickly abandoned their focus on recruiting young women to the factory floor. And none imagined that advocating for gender equality in the home to reduce women's household workload might be a necessary strategy for unionizing women in the workforce.[15] Instead, they shifted their attention to blocking the uptake of newer technologies that sped up the work pace. Some unions did manage to shut down piecework and press shop owners into creating better-paying, stable employment for skilled and unskilled workers. Invariably, trade unions' core membership and base of white, able-bodied men were the first—sometimes the only—group to find their way to less precarious work.[16]

Headlines of faulty machinery mangling children's limbs in meatpacking plants, textile workers trapped in shop-floor fires behind locked factory doors, and toxic fumes that enveloped workers mining phosphorus for matchsticks filled newspapers across the country for the first two decades of the 20th century. States enacted their own regulations to force companies to compensate injured workers. The patchwork of laws only highlighted how widespread a practice it was to neglect day workers in the industrial era. But it would take the widespread economic panic of the Great Depression in the United States to give labor organizers the public groundswell of support they needed to push for federal regulation of the workplace.

## INVENTING THE WEEKEND: WHEN FULL-TIME WORK IS PRIORITIZED, CONTRACT WORK PAYS THE PRICE

The National Labor Relations Act of 1935 cleared a route for workers to legally challenge employers. It was a first step in stopping the mounting standoffs between factory and mine owners who hired private, armed militias to threaten and beat workers and those workers who walked off shifts because of dangerous working conditions and came back ready for a fight. The act also created the National Labor Relations Board, to serve as a neutral third party to ensure that employers didn't interfere with employees' right to unionize or to bargain with management.

Importantly, the Wagner Act did not cover workers in several sectors, including agriculture and domestic service. It also did not apply to employees considered "supervisors" or those workers employed by federal, state, or local governments. And it excluded laborers considered independent contractors, such as domestic help, farm hands, and the relatives of small-business owners. The timing of the Wagner Act is key. It targeted the rapidly expanding, increasingly dangerous worlds of manufacturing and mining. By 1930, accidents at industry work sites had killed scores of workers across the country. The Wagner Act also became law at the height of the Great Depression, triggered by the 1929 stock market crash, when more than 15 million people, or just over 20 percent of the U.S. adult working population at the time, were unemployed and had no security beyond what their families could provide. The average citizen could see the cost of unfair work conditions on every street corner.

The momentum of the Wagner Act opened the doors to further reforms. It is fair to say that labor organizers advocating passage of the Fair Labor Standards Act (FLSA) in 1938 invented "the weekend," mandating, for the first time in a nation's history, a 40-hour workweek and guaranteeing a minimum wage — 25 cents an hour, or around $4.50 an hour in today's dollars. It also required employers to pay overtime to employees, at "time and a half" for hours worked beyond 40 hours a week. Minors, with exceptions for those working on farms or in the family store, could no longer be recruited to perform dangerous tasks. Prior to this legislation, children's small bodies made them the top recruits for some of the most unsafe work conditions. But a key element of the Fair Labor Standards Act also explains why so much contingent labor would not fall under this act, which would pave the way for expanding reliance on contract work.

The FLSA applied to "any individual employed by an employer." This certainly made sense, as the vast majority of economic expansion involved dangerous wage work generated by manufacturing assembly lines and extracting raw materials from the ground. Piecework, unfortunately, fell between the cracks of federal regulation. At the time of the FLSA's passage, organized labor had no interest in protecting piecework as independent contract work. Any work off-site or away from a shop floor could be used to pick away at work-site-based union membership or take work away from the majority of workers. If organized labor couldn't convince large numbers of women to leave home-based piecework behind and move to cities for union jobs, it would focus on the men who did.

Unions established their base in mining and manufacturing, after the passage of the National Labor Relations Act of 1935 and the Fair Labor Standards Act, three years later. At the time, close to 25 million U.S. workers belonged to a union. But efforts to expand to other industries stalled for the next decade in the face of World War II. The war effort diverted young men away from factories to the front lines of Europe. Unions felt that they had lost their core membership. Several unions brokered deals with employers and agreed to delay strikes and collective bargaining efforts until after the war. The famed image of Rosie the Riveter could easily stand in for the temp worker keeping the metal, transportation, and chemical industries alive during the war. She was expected to go back home once her brother or sweetheart came back from the war. As such, unions did not see a need to argue for expanding workers' rights until men returned to their roles as the primary breadwinners working full-time on the line.[17]

Once the war ended and manufacturing industries ramped up production lines for new products in glass, plastics, and metal, unions got back to work, too. By 1946, five million Americans took part in pitched mass strikes, coordinated boycotts of non-union shops, and unauthorized strikes across all manufacturing sectors.[18] Walter Reuther, president of the United Auto Workers (UAW), organized a series of postwar strikes, the first of which, in 1945, brought out more than 300,000 General Motors workers. It took another five years of massive strikes and negotiations with GM, Ford, and Chrysler, but by 1950 Reuther's so-called Treaty of Detroit had redefined what full-time employees might expect from their employers. Workers in the auto industry won a cost-of-living allowance (COLA) pinned to their annual wages, fully funded retirement pensions, social security contributions, and vacation time, as well as health and unemployment benefits.[19] In exchange, automakers won a five-year reprieve from annual strikes and—what would turn out to be most damning to future workforces — complete authority over production. Workers would no longer be able to question scheduling, tool redesigns, or manufacturing plant remodels aimed at breaking down full-time workers' tasks through automated processes.

The Treaty of Detroit was instrumental in tying retirement plans and health benefits to full-time employment in the United States. While none of these benefits came from a federal mandate that could provide this social safety net to other employees, workers' successes in Detroit gave employees in other industries renewed hope. But it also cleared the way for manufacturing companies to focus on reorganizing their factory floors in a way that moved machines in and people out, as quickly as innovation would allow. And, unfortunately, at the same time that the UAW was winning its battles for better employment terms, businesses in other sectors were organizing their own counterstrikes to challenge the union movement at large.

The Republican-controlled Congress pushed the Labor-Management Relations Act, drafted by Senator Robert A. Taft and Representative Fred A. Hartley Jr., through in 1947, overriding President Truman's veto of the legislation. Backed by the lobbying power of the National Association of Manufacturers, the Taft-Hartley Act took direct aim at the growing power of labor unions after World War II. Taft-Hartley, largely still in place today, banned "jurisdictional strikes" or work stoppages aimed at management decisions concerning workforce assignments. This made it much harder for unions to organize against employers using technologies to reduce a particular part of their workforce. And this, in turn, made it impossible for workers to consider, let alone voice, how loss of one set of co-workers might affect their own workflows.[20]

In addition to this wedge to organizing, work sites could no longer operate as "closed shops," meaning that companies could recruit new employees and build out lower management and professional staff positions, ineligible for union membership, to grow their non-union employee ranks. At the same time, in a reversal of the FLSA's provisions, the act allowed bosses to deliver anti-union messages in the workplace. As the ink on Taft-Hartley dried, conservatives in Congress rallied for what seemed like a small change to the National Labor Relations Act, considering the scope of Taft-Hartley. This tweak to the NLRA would prove to be as detrimental to the union effort among non-union workers as any blow delivered by Taft-Hartley, even though it seemed to narrowly take aim only at paperboys.

News magnate William Randolph Hearst lost a case in 1944 before the Supreme Court when he failed to convince the court that paperboys were the equivalent of their own bosses, operating as contract workers, exempt from employment labor protections under the Fair Labor Standards Act (FLSA). It's worth noting that paperboys didn't just deliver the news. They served as the primary sales force for papers, and their jobs often put them in physical danger as they made their way through the thick of bustling street traffic.[21] The Supreme Court read the spirit of the FLSA as applying to any worker economically dependent on and producing for another entity.[22] The Hearst Corporation, bruised from the Supreme Court's verdict recognizing street corner paperboys as workers deserving of workplace protections under FLSA, lobbied to narrow the definition of "employee" under the Taft-Hartley amendment to Wagner's National Labor Relations Act. Hearst pushed to explicitly exempt independent contractors, working off-site or considered peripheral to business operations. Congress's revisions to the original Wagner Act would instead require courts to use strict tests to classify "common law" employment status rather than assume that businesses hiring people to carry out work counted as workers deserving of fair labor practices and employment benefits.

The difference between someone hired as an employee and one who was commissioned to carry out independent work seemed clearer at the time. The dominant employment model of the day revolved around a physical workplace, 40 hours of shift work, and relationships between employers and loyal workers that might span 50 years. Bosses and employees might grow old together. But the Supreme Court's interpretation of the FLSA left open room to imagine a different work model.

The court held businesses accountable to those individuals working for them, no matter where they did that work or for how long. In the end, Hearst got the changes that he demanded. The door opened by the Supreme Court's 1944 decision to interpret the Fair Labor Standards Act as covering all workers was closed by the Taft-Hartley Act's explicit litmus tests and classifications for full-time employment and independent contractor work. This change also created a new class of non-unionized, expendable labor more valuable to the rapidly expanding service industry sectors and so-called information economy that was now booming in telecommunications, aerospace, and consumer retail. In sum, companies selling everything from advertising to space exploration could grow without investing in long-term, full-time employees.

## TEMPORARY COMPUTERS SENT US TO THE MOON

The erosion of labor protections for workers seen as "unskilled" but working outside of unionized manual jobs moved like water through cracks, exploiting and capitalizing on society's assumptions about whose work needed protection and who was worth protecting. Buried within the exemptions of the FLSA and the Taft-Hartley Act are clues about mid-20th-century assumptions about which workforces needed protection from unemployment and what—or who—seemed safe from automation. For example, the FLSA excluded volunteering, what we might think of today as the ubiquitous unpaid college internship. Volunteering was considered apprenticing, core to building one's professional identity. Since medieval times, divinity, medicine, and law, the so-called "learned professions," were seen as the exclusive paths of the educated classes. These skilled vocations needed no workplace protections. Their advanced degrees insulated them from economic insecurity. That is how a college education came to be seen as a gateway to the middle class for anyone who wanted to leave behind the mines or factories.

Most doctors, lawyers, and clergy operated as small businesses, typically self-employed, non-unionized professionals pooling resources through private practice. The work conditions of the professions, at least in the early days of modern industrial capitalism, did not seem relevant to the debates over FLSA. And so, the Fair Labor Standards Act applied to "any individual employed by an employer" but not to independent contractors or volunteers training to enter the professional class. Both types of workers were considered liminal characters. The contractor was treated like a hammer or mechanical pencil, there only to do an immediate task. Interns, by contrast, were being groomed to step into the corner office someday.

By the early 20th century, any occupation shaped by advanced training, certification, and professional codes of conduct was valued as a skilled profession. The de-skilled worker in a factory setting became synonymous with the unionized worker able to protect their position and workplace through federal regulations. Anyone considered "non-exempt" from the Fair Labor Standards Act could expect to receive overtime for hours worked beyond the maximum workday.[23] But tucked in the regulations were exceptions. Salaried clerical and administrative workers, though some of the lowest-paid employees, could be legally required to work beyond 40 hours and were not eligible for overtime. Professionals doing creative or non-routine work in the office were also exempt.[24] These exemptions help explain how independent, contingent workforces became a standing army of staffing and temp services, and its own lucrative industry. Much of the exempt contracted, temporary labor grew in the shadows, supporting professionals as they built out information services industries like accounting, scientific research, law, engineering, and finance at the close of World War II.[25] Educations in law and medicine remained socially coveted and the only gateways to highly skilled white-collar professions. But the expansion of administrative services of all kinds meant that the "learned professions" were no longer the only ways to find economic opportunity and social respectability.[26] They came with a price tag: job security.

Take, for example, the human computer pools at Langley Memorial Aeronautical Laboratory, in Langley, Virginia. Since the 1600s, the term "computer" has been used to describe a human doing hand calculations.[27] By 1946, thousands of young women had been recruited and trained to work as "human computers" for the U.S. Civil Service Commission at research centers like Langley Field all across the country.[28] They were the computational processors behind everything from deciphering coded messages sent by Nazi Germany to calculating how much thrust test rocket engines created and how to adjust their weights and heights to make them go faster.[29] Langley's first female computing pool opened in 1935. Under the FLSA, the federal government's War Department was granted latitude to employ independent contractors as civil servants.

By 1946, Langley Field's campus had reorganized as the National Advisory Committee for Aeronautics (NACA), NASA's precursor, with only a handful of women holding the professional title "mathematician."[30] The majority remained designated as lower-paid "subprofessionals" for several years after the war, including African American engineer Katherine Johnson, featured in the 2016 hit movie *Hidden Figures,* who would go on to calculate the launch windows for Apollo 13.[31] The lower wages and ranking attached to computers reduced lab costs. Directors could justify the entry-level, contract-based assignments because none of the women hired would be considered "professional" with no more than high school or vocational training. Johnson and fellow lead computer Dorothy Vaughan, who ran Langley's all-black "West Area Computers" wing, were hired at Grade P-1. This level guaranteed the women $2,000 per year, more than twice what they could hope to earn at their jobs as teachers in the segregated high schools of the South. But the terms of employment made it clear that they were not seen as the valued workers at Langley. They were hired "for such period of time as your service may be required, but not to extend beyond the duration of the present war and for six months thereafter."[32]

For those first few years, time off for contract workers was not an option. Even holidays were considered workdays as long as the U.S. space race against the Soviet Union was on.[33] Johnson and Vaughan eventually went on to professional rankings and full-time employee status. But the majority of computers were fired, as at-will contractors, once the predominantly male career engineers at Langley Field felt comfortable with the IBM 704. Still prone to overheating, the IBM 704 became the first computer able to reliably run calculations that had taken hundreds of women to confirm. It took nearly a decade after the arrival of the IBM mainframes at Langley for engineers to start fully turning over their work to the machines, keeping a pool of human computers on-site to double-check the machines' output. This made women like Johnson and Vaughan paradoxically contingent yet indispensable. Langley was not exceptional.

The human computers at NASA's Jet Propulsion Lab (JPL), in Pasadena, like the women at Langley Field, were recruited in the 1940s and '50s as contractors or temporary workers. They made all the calculations at JPL behind the country's earliest missile launches and bomber flights over the Pacific, as well as the United States' first satellite and guided lunar missions. They even gave the Mars rover, the planetary exploration vehicle sent to Mars, its first launch plan.[34] The majority of women at JPL had a similarly limited career path as the contract workers at Langley. Most started out as paid temps through the Works Progress Administration during World War II. JPL posted ads for part-time help in the mathematics and physics departments of surrounding colleges, calling out, in all caps, "COMPUTERS URGENTLY NEEDED."[35] Below the headline, the job description implied, clearly enough, that young women were its target audience: "Computers do not need advanced experience or degrees but should have an aptitude and interest in mathematics and computing machines."[36] In those days, if an office job didn't require an advanced degree, that meant it was open to women. Typically, the expectation was that women, as much as their white-collar employers, would not want a job to last more than the intervening years between graduating from high school (or, more rarely, college) and getting married. This was made even easier because of the lack of laws that prevented discrimination based on marital status. It was perfectly legal for managers at JPL, like those on the other side of the country at Langley, to fire women once they shared their plans to marry or became pregnant. Workers hired on contract, like human computers paid through federal government contracts, were not seen as valuable employees to cultivate. They were considered interchangeable, adding little to no skill, expected to leave when their projects ended.

Early forms of ghost work continued to flourish throughout the 20th century. Fast-forward to the 1980s, when temporary staffing agencies like Kelly Girl Service and Manpower contracted out more workers than most companies retained as full-time workers. Manpower's temp pool surpassed GM's full-time employees in the 1990s.[37] Sociologist Erin Hatton makes a compelling case that the wholesale swapping out of full-time employees for contingent staff was not just the fallout of technological change. She charts the staffing and temp agency sector's growth as its own service industry. What did it sell? The suggestion that young women and others, constrained by family care, limited educational opportunities, or geography, would reduce the companies' costs. Hatton argues that this "liability model," which treats workers as a drag on corporate profits, overtook the "asset model" of employment. It is possible that the asset model never made it out of the manufacturing sector after the passage of the Taft-Hartley Act in 1947. What is clear is that, by the time the internet was born, the U.S. was already moving quickly to a reliance on contract-driven services indefinitely staffed by contingent labor.

## THE NEXT WAVE OF EXPENDABLE WORK: OUTSOURCING AND PERMATEMPS

By the late 20th century, telecommunications and large mainframe computers had paved the way for the next iteration of piecework: *outsourcing*. Companies could move data-processing management, customer service, and even managing employees' records to any location in the world, as long as it was parked on a reliable node on the global communications network. Large multinationals based in the Global North, like British Airways, carved up and handed over chunks of their business operations to small firms in their English-speaking former colonies in the Global South. By doing so, they shed the obligations, costs, and worker safety nets that accompany traditional employment classifications. Nation-states like India sweetened the deal for conglomerates with state-funded expansion of global satellite systems and tax-free IT industrial parks. These new links in the global supply chain hired and fired locals at will to process everything from airline schedules and insurance audits to full-time employees' paychecks.

India's economic liberalization, in 1991, included the development of the Software Technology Parks of India (STPI) in every major city in the country.[38] Finance minister Manmohan Singh, a Cambridge- and Oxford-trained economist who would later be prime minister, argued that India's socialist economy, long propped up by its trading partner the Soviet Union, would need to liberalize and to deregulate its markets to survive the fall of the Iron Curtain. Key to his economic strategy was attracting as much foreign investment as possible. This is why, in Indian cities that lack free public utilities like tap water and sewer systems, there are high-speed broadband internet infrastructure and power grids to feed technology parks. The sharp growth in readily available internet connections, combined with a native English-speaking workforce and its long history of advanced educational training centers in science and engineering, made India the first epicenter of business process outsourcing (BPO).

But outsourcing was never simply about cost cutting. It was also about the growing resistance to unionization and evading long-standing labor regulations. As companies expanded their reliance on a far-flung network of contingent staff, they shrank the number of on-site, full-time employees who were eligible to collectively bargain or to push for increasing workers' benefits. Following a largely untested management theory, a wave of corporations in the 1980s cut anything that could be defined as "non-essential business operations" — from cleaning offices to debugging software programs — in order to impress stockholders with their true value, defined in terms of "return on investment" (in industry lingo, ROI) and "core competencies." David Weil, former administrator of the Wage and Hour Division of the U.S. Department of Labor, aptly described this process as the "fissuring of the workplace."[39] Everything that could be done by anyone, anywhere, for less money was now taskified and outsourced. Stockholders rewarded those corporations that were willing to use outsourcing to slash costs and reduce full-time-employee ranks. Yet, no matter how much streamlining happened, certain tasks, including auditing, filing, drafting, formatting, and shipping, still needed to be completed. By someone. Somewhere. The deep irony of outsourcing was that it was never restricted to contracting jobs to contingent workers far away. Outsourcing could happen through the cubicle down the hall.

## THE RISE OF THE PERMATEMP

In the late 1980s, Microsoft was thrust into the spotlight not so much for its status in the growing tech industry as for a troubling trend in its staffing procedures. Employing temporary workers and independent contractors was nothing new for tech companies. These workers were supposed to fill short-term company needs, pick up work when permanent staff were on leave, or provide expertise in areas outside of everyday operations. However, Microsoft was assigning temporary (or contingent) workers tasks that were virtually identical to what their permanent staff did. These "permatemps" spent years with the same responsibilities, reporting to the same management, and on full-time hours. By 1989 the IRS had grown wary of this arrangement and audited Microsoft's staffing procedures. The agency ended up deciding that about 600 of Microsoft's independent contractors should be reclassified as permanent employees, because their work was entirely under Microsoft's control.[40]

But it didn't end there. Microsoft outsourced the HR and payroll of its temporary workers to professional staffing agencies, firing workers who refused to be "converted" to temporary status. These agencies provided cheaper labor for Microsoft, as the company was not required to provide benefits or stock options for these permatemps now that they were officially employees of another company. This came to a head in 1992 when a group of temporary workers filed a class-action suit (*Vizcaino v. Microsoft*) against Microsoft claiming that they were common-law employees and should receive the same benefits as permanent staff. Microsoft argued that these workers had signed contracts indicating that they were temporary workers and not entitled to benefits and were compensated with "higher pay" and "flexibility." Over the next few years, Microsoft went to great lengths to differentiate temporary employees from its permanent staff, including different badge colors, different email addresses, lack of discounts at the company store, eliminating parking accessibility, exclusion from social events, and certainly none of the company's financial and medical benefits.

Finally, after nearly eight years of litigation, roughly 8,000 Microsoft permatemps received a settlement of $97 million. That may sound like a big win, but let's not ignore the loss. Namely, that the case was settled out of court. Without a court ruling, the question of what kind of worker these permatemps were and what kinds of protections they deserved has never been fully resolved. Yes, the permatemps case became a landmark example for legal and business leaders to illustrate how one should or should not implement temporary workers and independent contractors. But it did nothing to resolve the plight of the millions of people doing jobs that fall outside of the formal definitions of "full-time work."

## MURKY WATERS OF EMPLOYMENT CLASSIFICATION

As the ebb and flow of teenagers tending spinning jennys, human computers calculating moon shots, and call center operators in India staffing service calls suggest, technological advancement has always depended on expendable, temporary labor pools. Tracing the continuities rather than radical breaks from this past put ghost work in context. As history illustrates, hiring people on the assumption that they will be around only for the duration of a finite project or that the presumed efficiencies of an automated process can replace those workers is not radically new. By the late 1800s, textile mills in Lowell, Massachusetts, paid farm families to hand-fashion cloth pieces into shirt flourishes still too delicate to churn out on the factory floor. Similarly, today's companies perfecting search engine queries hire on-demand workers to test their latest ranking, relevance, and crawling algorithms.

Contingent work was further devalued by culturally loaded notions about what counted as a learned profession or "skillful" work, and which workers deserved or needed full-time jobs. Farm wives sewing, young black women tallying numbers by longhand, a continent of "others" doing data entry offshore, a contract worker helping with a speculative educational software package that may or may not ship could be written off, in part, because of their gender, skin color, nationality, professional training, physical location, or all of the above.

Those doing on-demand jobs today are the latest iteration of expendable ghost work. They are, on the one hand, necessary in the moment, but they are too easily devalued because the tasks that they do are typically dismissed as mundane or rote and the people often employed to do them carry no cultural clout. As the next chapters illustrate, on-demand workers live this fundamental contradiction while also signaling a fundamental shift.

Look closely at the realities of technology's shortcomings as they stretch across the industrial era. It is the paradox of automation's last mile—not automation itself—that episodically unsettles the meaning and value of human labor, provoking pitched battles over what types of employment matter and who deserves worker protections and investment. The move to postindustrial service economies has also sparked a second boom in what analysts call "knowledge work." In short, knowledge work is the conversion of the creative expertise required to think with and massage data into consumable services delivered online by industries from tech and law to finance and entertainment.

Thinkers from Karl Marx to Adam Smith imagined that machines played a critical role in "de-skilling" human labor. For Marx, automation dehumanized workers. For Smith, quickening the pace and expanding the reach of machines left a clearer picture of what was divinely unique about humans. Both men, products of their time, assumed that automation's intrinsic capacity to conquer all routine work was inevitable. This belief in the fundamental order and power of science to relieve humankind of its burdensome labor defines the Enlightenment and the industrial boom that followed it. Yet knowledge work produces a steady stream of always-on, intangible information services, from entertainment to tax advice. Neither the rise of practical physical services, like home repair, nor the boom in immaterial knowledge work, like text-based healthcare support, can be serviced by either a traditional assembly line or a 20-year career ladder. This world of work defies an economic ranking of "meaningful employment" along a skills ladder.

Early industrialists sought to divide labor between machines and full-time employees, but that division couldn't see, much less value, the workers who filled in the gap. Similarly, as tech engineers and businesses looked to automate productivity, they generated a demand for people to step in, for an indeterminate stretch, to do what economist Frank Levy and computer scientist Richard Murnane refer to as the "expert thinking" and "complex communication" required to make services work, as promised.[41] Caught in the paradox of automation's last mile, these workers are pointed toward some vague exit, attached to an imagined automated future floating on the horizon. A century later, contingent labor remains tucked in the folds of a busy production loop hidden from sight and impossible to fully credit or value.

Twentieth-century businesses automated production because they, like their contemporaries, believed in the power of technology to make life better.[42] Some also believed in the management mantra that workers were costly liabilities. With the growth of outsourcing and the expansion of temporary staffing, it became harder to believe that staffing cuts were limited to redundancies with machines. Soon everyone looked replaceable, reinforcing the tacit belief that the remaining full-time employees were the true assets.

Unions' tight focus on protection of full-time employees made sense through much of the 20th century. Historically, organizing full-time employees meant persuading those at the same work site to join forces and channel a unified voice, to set the bar for wages benefits and work conditions when facing managers and bosses at the bargaining table. The early days of organized labor and unions focused attention on protecting workers on the assembly line, in part because those lines and mines opened at the dawn of manufacturing were deadly business. Labor unions and widespread popular support made the United States one of the first countries to pass a series of comprehensive policies that brought about the security of our social contract, namely the stability of a 40-hour workweek, with time off; workplace safety and health mandates; social security and disability insurance; and the later benefits of employer-subsidized healthcare, retirement, sick leave, and vacation.

By the beginning of the 20th century, organized labor had focused its attention on championing the plight of employees working on an assembly line. There were good reasons for this. Employment conditions in factories and mines were appalling. Hours were long. There were no safety protections or disability insurance plans. Factory doors were routinely locked to keep workers from slipping out for breaks or leaving early. And, unlike the mostly domestic settings of hand "piecework" in the United States, factory floors could be observed and challenged by workers.[43] Just as important, factory assembly lines gave labor organizers a way to reach a membership base and to build a sustainable business model of association dues to fund its organizing work. If workers stood a chance against bosses more invested in profits than in worker safety, they had to collectively organize and bargain as a block. Workers' physical, full-time presence at factories made it much easier to build the on-site solidarity and mutual expectations that labor organizers needed to make strategies like strikes, slowdowns, and contract deliberations possible.

Some unions in the United States spent the better part of the last century recognizing the need to galvanize and advocate for not just women and communities of color but the service workers, domestic and home care assistants, and legions of others exempted from the federal regulations since the 1930s. But the attention to contingent labor arguably came too late, well after businesses selling lattes rather than steel had shifted to contingent staffing. Labor advocates have yet to iron out a strategy for building solidarity among workers who do not share a worksite or even a language for the type of work they do. Typically, those doing work on demand are blocked from interacting with each other, and just as often, as readers will see in the following chapter, they possess vastly different mental models for and investments in their identities as workers on demand.

Today's on-demand workers represent a fundamentally new type of worker. Their temporary status is part of what makes them indispensable. Their lack of fixed connection to any single company generates a common pool of workers available to a host of businesses able to tap this well of shared experience, availability, and diversity to constantly invent new projects. Companies now depend on the availability of this commons of contingent labor because they are constantly having to reinvent what they deliver to their customers, and they cannot meet this demand for 24/7 attention and rejiggering with a stable of full-time employees alone.

To be clear, today's reliance on contingent labor was not inevitable. Unions could have thrown their energy into organizing contract workers in both the private and public sectors. Courts and the federal government could have pushed more against the exemptions that allowed contract work to flourish outside of heavy manufacturing and mining. The U.S. Department of Labor could have classified work differently so that a more accurate head count of independent, alternative work arrangements was possible. And businesses could have been pressed by the general public to balance stock growth with profit distribution that benefited all workers, no matter the employment classification. None of these historical forces converged.

By the time the permatemp case settlement came about, in 2005, the global economy was no longer driven by building stuff. Though manufacturing remains a key site of employment, far more businesses profit from delivering services that create and manage customer experiences. This made distributing knowledge work via application programming interfaces, or APIs — also called "crowdsourcing" — a viable business model. These businesses now depend on assembling a diverse pool of workers to tackle projects, relying on semi-automated systems to move work assignments to and from each other's desktops.

With the networked technologies and staffing set up through outsourcing and vendor management systems, incumbents and startups could hire contingent labor to spin up projects quickly and rapidly iterate on a prototype or beta version of a consumer offering, assuming that these workers would stay in the picture only as long as the project deadline. It's worth noting that some of the largest "full-time employers" are actually staffing and temp services like Accenture, and some of these companies now rank among the top job providers in the world.[44] Of the top 20 global employers in 2017, five are outsourcing and "workforce solutions" companies, according to an analysis by S&P Global Market Intelligence.[45] For comparison, in 2000, only one employer in the top 20, International Business Machines Corp. (IBM), offered outsourced IT services. The reason it is so hard to tell how many legions of temp staff now stock businesses with the personnel they need is that these intermediary staffing agencies are the very ones carrying — or covering up — businesses' labor costs. And it shows.

The outsourcing sector has boomed during the past two decades. Between 2000 and 2016, the annual value of outsourcing contracts tripled, from $12.5 billion to just over $37 billion.[46] Thanks partly to double-digit growth in big technology projects as more companies transfer massive volumes of data to the cloud, the outsourcing market is expected to rise again in 2017 and 2018.

Given the mechanics of outsourcing and vendor-managed staffing companies, the cost of procuring temporary staff is accounted for in literally the same way as copy paper in quarterly reports to stockholders. When businesses announce that they're reducing employee ranks, stock listings go through the roof. But as the permatemp case illustrated, it is not uncommon for companies to swell their temporary staffing budgets within weeks of layoffs. They "buy back" the labor of former employees through temporary staffing agencies to make it easier to close out both projects and labor costs at the same time. And younger companies, including numerous Silicon Valley startups, used to relying on contractors for everything from beta testing to viral marketing, could maintain their "lean" look on paper while employing the services of an equal number of staffing-agency-contracted temp workers to do what they do. Relying on outsourcing and contracting workers through staffing agencies as the "employers of record" primed the so-called platform economy.

Employment practices that convert workers into something "procured" to launch a product are baked into ghost work's code. Over the past decade, companies launched on-demand services that matched individual consumers or other businesses to everything from a ride to the airport to medical transcription through the touch of a smartphone app. They sell themselves as hip and innovative technologies, not sites of employment or temp agencies. But they and the scores of on-demand ghost work platforms like them, intentionally or not, are quietly taking over the $115 billion temporary staffing industry. The platforms detailed in this book, offering direct access to a pool of contingent workers online, are not outliers.

The consumer appetite for new experiences means that most businesses work to position their new goods and services as the latest, greatest thing. That means that they are constantly putting together projects that have immediate needs or are speculative. Every worker is necessary in the moment but then expendable once the project ships.

# PART II

## Demanding Work

# 3

# Algorithmic Cruelty and the Hidden Costs of Ghost Work

## Indifferent Design and Its Unintended Consequences

Most on-demand workers accept an uneven workflow as part of the API landscape, but software bugs can make the work feel even more precarious. In 2013, a glitch in the system caused Joan's MTurk account to be abruptly suspended — an on-demand worker's worst nightmare. "I never received any emails notifying me of a software issue; I just knew I couldn't log in anymore," she says. "I called customer service and was told I had to wait until the site fixed the issue. The suspension cost me close to $200. I lost high-paying work because of a problem on their end, not because of the quality of my work." On-demand workers like Joan have no way of knowing what's happening, much less recourse when things go awry. After 40 hours of being locked out, Joan got her account back. But the experience left her wary. "It was 40 hours of not knowing whether or not I was going to be able to do my job or maintain my income, and all for reasons that were not disclosed to me."

In 2014, Eric Meyer, a web design consultant and author, coined the phrase "inadvertent algorithmic cruelty" to describe a flaw in computational design — a lack of empathy.[1] It poignantly captures Joan's experiences with ghost work. Meyer first used the phrase in a blog post in response to Facebook rolling out its "Year in Review" feature, which shows people highlights of their year in pictures. In Meyer's case, the app worked as designed—it showed him pictures from his past year. The problem was that his five-year-old daughter, Rebecca, had died of brain cancer that year. Later, in an essay for Slate.com, Meyer wrote, "A picture of my daughter, who is dead. Who died this year. Yes, my year looked like that. True enough. My year looked like the now-absent face of my little girl." He acknowledged it wasn't a deliberate attack, just an unfortunate result of code. "Algorithms are essentially thoughtless," he wrote. "They model certain decision flows, but once you run them, no more thought occurs."

When the design of an algorithm, platform, or API lacks thought and is unleashed on unsuspecting consumers, people like Eric Meyer suffer the unintended consequences. When "thoughtless processes" are introduced into the workplace, especially one with low-income earners who have little bargaining power and with a lot to lose, the unintended economic and social consequences are severe.[2]

Ghost work markets do not make the transaction costs associated with getting work done evaporate. Instead they shift those costs to on-demand workers and requesters. While software can be fixed, the bigger issue is a system that turns a blind eye to workers when things break down. As of today, on-demand platform companies and those using them to hire workers have no terms and conditions that compel them to be accountable to workers. Even the payment obligations of this new form of employment are left to glitch-prone software.

More specifically, workers absorb the cost of searching for work, learning how to do tasks, and communicating when things fall apart. Requesters absorb the cost of finding talent, building trust, and maintaining accountability with the workers. These transaction costs, which disproportionately fall on the workers, arise due to inadvertent algorithmic cruelty in the design of the platforms and their APIs. Software systems, used by both platforms and requesters, are too rigid and unforgiving to deal fairly with all of the complexity involved in hiring workers, evaluating their work, and paying them. So the humans, on both sides of the market, are left with the task of resolving these complexities at their own expense, though the workers bear the heavier brunt of these costs.

## The Cost of Doing Business

At the heart of the on-demand economy is the premise that relying on ghost work cuts transaction costs and, therefore, boosts profits. Transaction costs are those expenses associated with managing the production and exchange of goods or services. Nobel laureate Ronald Coase, a key contributor to modern economic theory, popularized the notion of transaction costs, though he did not coin the phrase itself. His seminal 1937 article "The Nature of the Firm" was published only two years after Wagner passed the National Labor Relations Act. In it, Coase argued that businesses had to coordinate their operations, such as finding, hiring, and training workers, to reduce market frictions. The only route to lower costs and to turning a profit hinged on making businesses run as smoothly as possible. In essence, he was the first economist to theorize how to produce at scale through modern private enterprise and profit from a well-oiled org chart.

Ghost work economies sell themselves as software that can eliminate the expensive frictions of searching, matching, training, communicating with, and retaining workers. Yet, as Coase might have warned, communication and coordination among workers, and between workers and their employers, not only is necessary but is actually money well spent. For all the claims that ghost work can combine algorithms, artificial intelligence, and platform interfaces to replace the company's function as "the entrepreneur-coordinator, who directs production,"[3] there is evidence to the contrary. The transaction costs of ghost work don't melt away. Instead they are shifted to the shoulders of requesters and workers. Requesters must juggle all the management that typically comes with scoping a new project and handing it to a new employee. They spend extra time and energy explaining tasks that they thought needed no explication once converted to code and relayed via APIs. Workers pay a disproportionately higher price: they lose their time, even their paychecks, with no opportunity to appeal any mistreatment. Many of the transaction costs passed on to requesters mirror those shouldered by workers. Each hurdle faced demonstrates that ghost work isn't working smoothly for anyone involved.

## REQUESTING WORK: TRANSACTION COSTS

The most commonly reported requester difficulty occurred in the process of matching a worker to a task. Requesters reported posting a task and then receiving a flood of applicants they would then have to sift through and vet, especially for larger, more complex tasks. A communications specialist at a PR firm said, "It's overwhelming how many people you get, because those people are very eager to do the work and it can be difficult to tell if they can actually do the projects." This is the flip side of workers being hypervigilant, as constantly being alert for good work also results in requesters getting flooded with applicants, which makes picking a worker hard. Another marketing manager said, "Especially when you're getting responses from around the globe, sifting through all those people can be painful." Vetting workers was a time-consuming, manual process that often involved a phone or Skype call. A VP of communications at a startup discussed the importance of getting the vetting done right: "A lot of the workers have good technical skills but poor communication skills, so you have to really vet for this. I've also had situations where the website they created might look good, but then you have another person come in later to do work on it and they find dirty code and unfinished code. So now I do more due diligence in selecting workers."

Some platforms provide guidance in the form of ratings or a reputation score, as described in chapter 1. Requesters used this information to varying degrees. Some reported using them, others took them with a grain of salt, and yet others ignored them completely, thinking they could be manipulated, since there is usually no verification or validation. Another issue with ratings and reputation scores is that they don't transfer from platform to platform. A worker could have a stellar rating on one platform but, when they move to another, have to start from scratch. Many requesters reported that ratings and reputation scores are not as good as portfolios of prior work and lamented that platforms make it difficult to see prior work. A marketing manager at a previously mentioned engineering firm crystallized this feeling: "I would say that [the] profile of the freelancer should be more open. I understand that the companies they have worked for won't really allow freelancers to display the work they have done for them, but details around what was delivered would certainly help." In general, requesters use on-demand platforms to source workers but found vetting them to be time-consuming. As a result, we will see the requesters often try to do this as few times as possible and reuse workers they've already vetted. Vetting workers, a transaction cost that, in the staffing agency setting, is absorbed by the agency, gets passed on to the requesters in the on-demand platform model.

After a requester chooses a worker, the next step in the process is actually getting the work done, and here, too, we see a number of transaction costs from the requester perspective. A number of these costs arise because the on-demand worker, once hired, temporarily becomes part of the firm until the work is completed. They are essentially a new employee, with no training, on their first day, assigned to do a task remotely. The cost of training is borne by both requesters and workers. But if workers do subpar work, it gets rejected and they don't get paid. If requesters get subpar work, they can simply choose not to pay and find another worker.

One of the first issues that comes up is establishing a relationship based on trust and accountability in this setting. It takes time to build trust between two people, but this is at odds with one of the major draws of using on-demand workers—to gain access to labor quickly. The rating and reputation systems that platforms implement, described above, are an attempt to convey trust through an API, but these are incomplete solutions. A procurement manager at an industrial supply firm said, "There's always a trust issue. We either trust they actually have the qualification they say they do and we just bring them on, or, if it's a more critical task, we'll bring them in for a 10- or 15-minute conversation first. I mean, this is all we can really do; at some point you have to have faith in a person's professionalism."

These trust issues can spill over into accountability issues. In a traditional staffing-agency model, if a worker does labor that is below the hiring firm's standards, the agency can be held accountable. But in an on-demand setting, there is little accountability. An online retailer who sells gifts and curios said, "We had nine rounds of recruitment firms and didn't find the languages that we needed or coders. Then we finally get this guy, and he was a nice guy and he knew the stuff. And we were just gearing up and he just ghosted." The retailer was helpless once the worker disappeared. In addition, if a relationship with a requester sours, an on-demand worker can restart with a new requester, which further erodes any sense of accountability on the worker's part. The marketing manager for the online education company said that freelancers "have a casual attitude. When we work with [full-time employees], they know they're being judged and appraised. When it comes to freelancers, for them it is one project and then they will have another project with some other company. They carry that type of attitude, which actually brings a lot of the trust issues to the table."

The benefit of allowing any worker to work for any requester, which is a design decision that most platforms make, has the unintended consequence of eroding accountability. The lack of accountability brings about another transaction cost borne by requesters in that they have to manage, and even micromanage, workers. Another marketing manager reacted to the lack of accountability by monitoring the workers closely: "I ask for daily updates from freelancers. They work remotely, so managing them is tough, but having them send a daily update helps. By getting the daily update, at least I know beforehand a deadline might be missed and I can put a contingency plan in place."

It can be hard for on-demand workers, who usually work remotely, to observe, absorb, and act according to the culture of the hiring firm. For example, the engineering firm marketing manager said, "I think the major pain point is that, since these people are not from our organization, they are mostly not aware about our timelines, procedures, guidelines, and way of writing. Every company has a different culture, and it becomes really difficult for freelancers to follow a single culture, since they might be working with five different companies with five different cultures and guidelines." Corporate culture can affect the expectations that requesters have for the work submitted and, in turn, affect the actual product produced by workers. Not being able or willing to absorb the culture of the hiring firm can result in on-demand workers submitting work that is not appropriate for the company, as the marketing manager found out when working with a designer: "He used to deliver good work, but our company had a problem with his bright colors. He was advised to use more blue, gray, and darker shades, but instead he used vibrant colors . . . Even after explaining to him many times our way of doing work, he just couldn't abide by company standards." Another marketing manager, in advertising, said that explaining the culture, something that is difficult to build into an API, to an on-demand worker is yet another transaction cost borne by the requester: "I can't expect a freelancer to have a firm grasp on our culture — that's not the nature of a freelancer — so it's really up to me to make sure their output meets company standards."

Trust, accountability, and culture are social aspects of the general work environment. One of the main technical challenges involved in hiring on-demand workers is giving them access to the necessary tools and data. Requesters overwhelmingly expected workers to have their own software tools to bring to the job. A communications specialist at a direct mail company said, "Yes, we assume they can do the job they're hired for, and if that requires certain tools, then, yes, we assume they have those tools. That's part of being a freelancer: you're independent. We aren't going to provide a freelancer with tools or train them; we're going with a freelancer so we can avoid making those investments." A full-time employee would be given all the software tools necessary to do his or her job. But in the on-demand labor setting, this cost is transferred to the workers. A manager at a consulting company said, "Imagine somebody saying that he can draw the architecture but doesn't have AutoCAD. That's not helpful at all, so yes, we need to make sure they have the necessary tools to get the job done. If they don't, then we find someone who does." Requesters can improve their bottom line, since they don't have to provide software tools for on-demand workers.

Requesters overwhelmingly used the same technique, maintaining their own trusted pool of workers off the platform, to overcome, or at least mitigate, these difficulties. Requesters build this trusted pool by repeatedly hiring and trying out workers.[4] First, requesters come to on-demand platforms to find and source workers. They take the time and effort to vet them, and those who pass are given a task. After the task is completed, those workers who performed well are added to the hiring firm's internal database of trusted workers. Then, when it comes time to hire another on-demand worker, firms start by looking in their trusted pool.

From a requester perspective, there are quite a few benefits to this approach. First, it saves money. Requesters meet workers on a platform, then, if they are good, they take the relationship, and future jobs, off the platform. Since platforms usually charge the requester some percentage of the fee paid to workers, requesters can save money by circumventing the platform, which improves the requester's bottom line. Second, hiring from a trusted pool allows a requester to vet workers and to show them the corporate culture and expectations just once and then recoup the cost of doing so by repeatedly hiring from the pool as the need comes up. Third, hiring from a trusted pool mitigates the risk that requesters face when hiring an on-demand worker. There is less chance of an incompatibility due to personality or work style. Fourth, if a worker has been hired before, he or she can build, or at least begin to build, a relationship of trust with the requester. In general, requesters feel that it's important and valuable to build relationships with the workers, as doing so leads to better outcomes.

Perhaps because requesters found a relatively straightforward way to mitigate the transaction costs passed on to them, they overwhelmingly reported that they would recommend using on-demand workers to others. The advertising firm marketing manager went so far as to say, "I would rate it an 8 out of 10. The quality of work that they deliver and the skills they bring with them makes a great impact on the project meeting the client's expectations." On the other side of the market, it's doubtful that workers would rate ghost work "an 8 out of 10."

## Ghost Work's Hidden Pain Scale

The burden of transaction costs used to fall on companies, but now it falls squarely on the people doing ghost work. Imagine a pain scale of the sort found in a doctor's office, with a smiley face at the low end and a wail at the top end. Algorithmic cruelties inflict pain on workers that can be mapped onto the scale. In some cases, the burden is a mere annoyance, a minor paper cut. Those little cuts may look like time lost to seeking work and understanding the work. But other times, small annoyances can fester into a more painful situation, one that becomes a drain on a worker's time and energy. This level on the pain scale may look like executing a job without any feedback or in isolation from peers and colleagues. What tops the pain scale for most workers is the risk of not getting paid. Most workers have zero recourse if something happens to their account and they aren't paid for their work.

The harsh irony is that ghost work platforms and individual requesters wash their hands of the pain they inflict on workers. Companies, from MTurk to Uber — and this is key — view workers as mere customers who are selling their labor, as they might sell their used record collection or rent a spare bedroom. In the eyes of ghost work companies, customers come to their sites strictly of their own volition. And, as customers, they can leave at any time.

Again, from a company perspective, workers are costs and liabilities.

Customers are free agents, buying and selling at their own risk. Yet when workers are not acknowledged as the essential engine driving commerce between ghost work platforms and requesters, it's the workers who suffer the most. As a direct result, the status of a workforce of millions is left in limbo.

## 1 TO 3 ON THE PAIN SCALE: HYPERVIGILANCE SPUN AS FLEXIBILITY

The ability to see good work, on a reputable platform, and pluck it from a stream of projects hinges on a worker's capacity to be in the right place at the right time. These workers must cultivate a hypervigilance that would have made Coase's head spin. In interviewing dozens of on-demand workers, we saw two types of hypervigilance. The first was the need to spend hours wading and sorting through spam or suspicious offers for "at-home work," searching for legitimate work on a legitimate platform. Because there are no legal requirements screening who posts work to on-demand marketplaces, workers had to make sure they weren't signing up to a site that was simply looking to harvest their email address or that would open them up to identity theft.

Lijo, 24, works for a business processing organization and lives in Bangalore, India. He learned about MTurk from a flyer stapled to a tree. He called the number, and the person who answered told him an account on the platform cost 1,000 rupees (almost $14). Unaware that registration on MTurk was actually free, Lijo negotiated the price down to 600 rupees ($8.25) — all he could afford at that time — and bought an account. The person explained the basics of how to use MTurk, but once Lijo was on his own, he quickly got confused. A year later he'd earned about $20 (roughly 1,450 rupees). "MTurk was a terrible investment of time," he says. "I didn't learn anything and risked sinking time into work that had no security. There is no person, no office, no one to answer questions." Since platforms do not vet the requesters or their tasks, workers are left with the problem of finding a reputable platform with work that is worthwhile on their own.

Many of India's on-demand workers are wary of unscrupulous companies. Most of the country's workers have a personal or family experience of being bamboozled by a fly-by-night contracting agency that made big promises but fell short in providing paid work. For instance, in the 1990s, a spate of call centers opened in the country. The companies would hire Indian workers and then disappear three weeks later, leaving unfulfilled promises of paychecks in their wake. That experience has left many young people like Lijo suspicious of online jobs that may be phishing for information or trying to fool them into working for free.

Pundits both champion and criticize the flexibility in these labor markets. Champions see flexibility as the saving grace of the new economy, while critics curse flexibility as a source of downward pressure on wages in the sector. Again and again, those doing ghost work are told what an amazing perk it is that they can work anytime, anywhere.[5] But more often than not, this so-called perk masks the reality of online work. The most hypervigilant workers, those always looking for the next task, are the most rewarded. In practice, on-demand requesters and the algorithms behind labor platforms automatically generate quick-turnaround deadlines for jobs, even when the work isn't time sensitive or a worker's availability — beyond turning an app on or off — could be added to the decision-making mix. That artificially imposed time squeeze means workers must constantly scan for jobs, especially if they want lucrative opportunities, which get snapped up quickly. In reality, flexibility is a myth. Instead of the utopian vision of an endless stream of online work that a person can dip into between other pursuits, on-demand labor more closely resembles the infamous *I Love Lucy* television comedy sketch with Lucy and Ethel working on the assembly line at a chocolate factory. As they scramble to keep pace, the pace of work comes faster and faster.[6]

The second type of hypervigilance we saw was the need for workers to be on call day and night. Job posters and platform designers assume that workers aren't navigating other time constraints. But we found that workers had to constantly search for job opportunities they could fit in around other commitments. One example was Natalie, an African American woman who lives with her parents in Queens, New York. The 27-year-old has a bachelor's degree. She likes doing on-demand work from her parents' home because, as an aspiring musician, she can balance her work and her art. But she struggles to budget her time, because the online work is unpredictable. Soon after signing up for one of the platforms in our study, Natalie discovered she had to constantly check project boards for work if she hoped to make any money. "The first worker or workers to respond to a post get the job assigned to them automatically, which was a little frustrating. It made me feel like 'Oh, I don't know if I'm actually going to be able to get work.'" Since platforms only allow workers to pick up tasks on a first-come, first-served basis, workers keep one eye on the platform in case a good task comes up.

In some cases, especially with the larger platforms, companies expect workers to stay logged in to forums, an informal gathering space for sharing information. Diane F., 59, lives near Washington, D.C. She studied biomedical engineering and worked in the field right out of college. But she switched careers early on and spent many years working with computers. Getting laid off from her job at a local university prompted her to look for on-demand work on LeadGenius. "I got unemployment for six months, then I decided perhaps I should look at doing something on my own."

Diane found LeadGenius dizzying at first. "I saw that we [the workers] had to be on HipChat." HipChat is a feature of LeadGenius's platform, an internal chat software. Some workers like the informal conversation in HipChat, but for Diane it was so distracting that she logged herself out. "I would get on HipChat and people were sending smiley faces and talking about what they ate for breakfast, and I've got stuff to do," she says. What she didn't realize was that staying on HipChat was a key job requirement. "They [LeadGenius] expect you to sign in and to be there, even if there's no work." Just before our interview, Diane's account was suspended. One of the reasons, according to the email she received, was that she hadn't logged on to HipChat in more than three weeks. She says, "I thought we would go onto HipChat when we wanted work, not that I had to be on HipChat all the time."

Our research shows that workers take advantage of chances to control their time.[7] We paid MTurk workers to categorize a batch of Amazon product reviews as positive or negative, which could be done in less than 30 seconds each. One-third of the workers had to do each categorization within one minute (no flexibility), one-third had to do each categorization within one hour (some flexibility), and one-third had to do each categorization within one day (lots of flexibility). Workers in each group were given the same number of tasks, and workers were randomly assigned to each group. When given more time to finish a task and gain more flexibility over their schedules, MTurk workers took it.

Recall that MTurk workers can accept tasks and then hold them in their queue to work on at a later time. On average, workers with an extra hour on the clock would start work 17 seconds after accepting it. But workers with a day to turn around tasks would start work slightly more than four minutes after accepting it. At first glance, the difference between 17 seconds and four minutes might not seem that large, but imagine how you'd feel if the longest you could step away from your computer was 17 seconds. Four minutes gives workers like Natalie and Diane the chance to answer a phone call, check on a napping newborn, go to the bathroom, or focus on other, more pressing work and not worry that this job will be taken by another worker. Furthermore, as we gave workers a longer amount of time to complete the task, once workers started the categorizing, they would do more work before taking a break and take a shorter break once they did. Workers, when given more control of their time, spend it wisely, taking breaks as needed and then sitting down and cranking it out when the time fits their schedules. "Flexibility" is an empty euphemism unless workers are able to set and control their schedules on their own terms.

Giving workers some flexibility changes how they spend their time. But how much do workers value flexibility in terms of dollars?[8] We measured the *compensating differential* — how much more a requester has to pay to get the same amount of work done when they constrain a worker's time instead of giving them control over their break times — to put a price tag on flexibility. Workers who had worked 0–10 hours in the previous week did not value flexibility as far as we could detect. But workers who did 11–30 hours of ghost work in the previous week valued flexibility so much that they would trade a requester $0.98/hour to get control of their time. And those who contributed the most, by doing more than 30 hours of MTurk work in the previous week, valued the flexibility at $2.37/hour. Similarly, workers who had not set a daily income target for themselves did not value flexibility as far as we could detect, and those who did valued flexibility at $0.92/hour.

Workers who set daily income targets or do more work quite likely need the money more, and these are exactly the workers who value flexibility the most. They already know how to make the most of their time spent on ghost work. Ultimately, the only workers who can afford to be less hypervigilant are those who don't rely on on-demand labor as their primary source of income. In other words, those who need the money the least have the most flexibility. However, if a worker relies on that income, the reality is that he or she can't afford to look away.

## 4 TO 6 ON THE PAIN SCALE: ISOLATION AND LACK OF GUIDANCE SPUN AS AUTONOMY

In most cases, people doing ghost work are left to their own devices to navigate the ins and outs of a job. In rare cases are requesters available to answer questions. And indeed, asking a question may result in a delay that costs a worker their job. Instead of risking it, workers dive in and lean on their own ingenuity. Most figure it out as they go, orienting themselves to the task at hand, figuring out what's being asked of them, and learning how to operate clunky, outdated interfaces all while getting the job done as fast and accurately as possible. But working without guidance and in true isolation takes a toll.

At 19, Ayesha was one of the youngest workers we interviewed. We met in her family's house in Hyderabad, India. She is the middle daughter of a joint family with 16 members. Her brothers already worked on MTurk, so they helped her open an account. She is completing grade 12. Ayesha's parents adhere to Wahhabist interpretations of Muslim sharia codes of conduct that, at times, conflict with Ayesha's own interests, like studying to become a doctor and having her own public clinic someday. But Ayesha's mother does feel strongly that girls should work and that working from home is good for girls.

Ayesha told us she feels afraid to work on the MTurk site by herself, because she might make a mistake. Even the smallest error can lead to an account getting blocked, so the stakes feel too high to warrant the risk. As a result, she goes to the site only when her brother is at home, but he works outside the house, so she's unable to work as much as she'd like. "I don't work much, because I am scared." For now, the money, along with any mistakes that Ayesha makes, flow to her brother's account. She hopes to improve her English and computer skills but doesn't see the site as a way to do more than contribute a small amount of income to help cover household expenses.

Even workers who have a good sense of how to do a specific type of project still work in a vacuum, with no feedback to help them gauge whether they are a good fit for the project other than the initial description of the task posted to the site. Workers don't know if they are technically or culturally competent for a task until they try to complete it. But tackling a job without knowing if they can succeed often presents a new risk. If workers discover too late that they are in over their head, their reputation is at stake, because workers' reputations are linked to tallies of their approval ratings. In other words, if their reputation gets dinged, future job opportunities dry up.

As we described in chapter 1, most platforms rate workers, in much the same way customers give businesses a Yelp rating, based on feedback from those doing the hiring. Since the platform does not provide any way for workers to get clarification or training, a worker who can't problem-solve on their own risks getting a low rating. For example, one worker we interviewed accepted a large batch of tasks from a requester new to MTurk who was asking workers to read one-sentence product reviews and then rate each product as "useful" or "not useful." The worker dug into the job, reading through pages of one-line reviews and ticking the box that fit his sense of whether each product was worth buying. The worker submitted the tasks for payment hours later, but the requester rejected them all.

Suddenly thousands of that worker's completed tasks were rejected, without explanation. When the worker looked up the same requester's reposting of the task, he noticed that the requester's latest version included one word that had been changed in the instructions. Either this was a completely different task or the requester had added more clarification to what he wanted—*evaluate the reviews, not the products*—and thus the worker's output no longer matched the job. Once the worker contacted the requester and pointed out the change, the requester did pay the worker for the completed tasks. But, to add to the irony, the requester never changed the status of the tasks from "rejected" to "accepted" (perhaps he didn't know how) and the worker's rating on the site plummeted.

To regain his reputation as a reliable worker, he had to effectively drown out the rejected jobs by accepting tens of thousands of penny tasks and doing them all perfectly. Only then did his reputation for accepted tasks return to its high standing. This is just one of the invisible ways the system advantages requesters over workers. Requesters are seen as visible and valuable customers to the platform, giving them latitude to change a job request midstream without censure. Those performing ghost work are mostly invisible and treated as interchangeable.

They must scramble to make do with a system designed to assume that workers could be adversaries trying to game or rip off the customer-requester rather than real people at work, doing their best.

Justin, whom we briefly met in the introduction, where he was doing image-tagging work, had previously had a good job at an upscale grocery store. When his wife landed her dream job in a new city, they weighed the costs of Justin relocating to another store location. The commute would have added hours to his workday. Day care for their two young sons also seemed beyond their means, even with two paychecks. Justin decided to be a stay-at-home dad and try his hand at on-demand work. He went online searching for "work from home," and MTurk popped up. He'd been using the platform for only two weeks when he saw our call for interview participants on the site and offered to speak with us about his experience.

What frustrated Justin most was seeing some requesters use what he called "a bait-and-switch strategy" to entice workers to accept tasks. He described a recent job he found, where the requester wanted the worker to type up handwritten notes. The example of the work that accompanied the post showed clear notes, but when he accepted the job, the notes assigned to him were, in his words, either "just gibberish or blurred out." In some cases, he says, "you can return the HIT [job] without having it held against you." But in other cases, he had no way to know that he was in over his head until it was too late.

That's what happened the time he took a job identifying features of furniture, such as whether a couch was a camelback or not. Requesters, understandably, keep task previews short, since it's difficult to gauge how much might be too much information for any given worker assessing a job. That means that previews rarely include detailed descriptions or varied examples of the project. Justin had to gamble with his time and hope that he'd chosen his assignment wisely. In hindsight, he didn't have enough information to know that he'd made a bad call taking on the furniture classification task. Justin had to spend a significant amount of time scouring the internet for images associated with the terms flashed in front of him before he felt he could correctly answer the questions associated with the task. "That ends up taking far more time than [the task] really needs." This type of stumbling block highlights a common cultural disconnect seen among U.S. and Indian workers alike. In this case, Justin's class and work background hadn't infused him with the cultural literacy he needed to identify and define terms associated with upscale furniture purchases. Similarly, we met workers in India hindered by jobs involving things like describing panini makers, a kitchen appliance not commonly found in Indian households.

And, unlike the job of typing up handwritten notes, which Justin could quickly determine would be a time suck, he took the job identifying furniture, thinking he had the smarts or the intuitive wherewithal to do a good job. But without feedback from the job requester, he had no way to know if his work was meeting expectations, until it didn't. By then it was too late to return the task, redo the work, or avoid having his reputation score dinged for subpar work. "That was the first time I had a job rejected. The job wasn't as easy for me to complete as I thought it was going to be," he says. "I didn't email them about why my work was rejected or challenge the decision. In the end, it's not really worth it." Justin would have to invest even more time trying to pin down what he'd done wrong, with no guarantee that anyone would ever reply to his email. He chose to cut his losses and move on.

In the absence of clear instructions or feedback on work under way, workers like Justin have no way to gauge how to succeed at any given job. Workers learning a new task or mastering a new platform compound this vulnerability. Due to the legal liabilities associated with "curating a workforce," as was discussed in chapter 2, those employing workers can't provide on-the-job training. As a result, those doing ghost work bear the costs of learning how to navigate not only the task but also the work culture, and they have no chance to ask questions or get feedback from requesters.

## 7 TO 10 ON THE PAIN SCALE: NOT GETTING PAID WHEN TECHNICAL FAILURE OR HUSTLE IS SPUN AS MALFEASANCE

Even after a person has cleared the hurdles of finding good ghost work, learning the platform's quirks, and completing the task, they still run the very real risk that they won't be paid for their work. Many payment glitches can be traced to errors in the platform design. People who design on-demand job sites assume workers have fast broadband connections and reliable power sources. In reality, millions of people doing ghost work do their jobs with outdated computers, faulty internet access, and even shared IP addresses. In the chasm between the platform designers' mental image of who is using their product and the reality lies a minefield of potential errors. One wrong step can totally blow up a worker's chance of getting paid.

A change of address, for example, can flag an account as being suspicious. Moshin, 24, lives with his mother in the southern Indian city of Kochi; his father passed away two years ago. He has two sisters who are married and live in the United States. He is studying for a master's degree in computer applications and works part-time on MTurk but feels deeply ambivalent about it since his MTurk account was suspended with no warning and for no reason. He suspects the suspension was triggered when he changed the mailing information for his paychecks. He explains that he felt better having the check sent directly to his uncle, who works in a posh neighborhood, than to his own address, in a makeshift location without reliable postal service. But changing the name and mailing address for his checks meant it no longer synced with the national identity papers he'd used to open the account. He suspects the switch initiated an automatic suspension. Yet, as happened to so many workers we interviewed, Moshin never received any official explanation for the suspension of his account, beyond a vague statement that he'd violated the terms of agreement. The terms of agreement is a multi-page, single-spaced, tiny print that all workers "accept" when they activate an account.

That kind of confusion was rampant among workers we met. Latonia, 28, an African American woman in Atlanta, has an associate's degree in graphic design and moved closer to her family after her father died. She was drawn to ghost work because she likes having several income streams. "I don't like having just one job." But when her mother's internet service was turned off, Latonia got a warning message from her manager on the platform. "That warning kept me from getting certain jobs." Then Latonia's phone, which was her primary conduit to the internet, was stolen. "I couldn't do my work on the phone, and I told my manager that, as soon as I could get back online [I would get to work], but she said, 'You still had to give me a warning,' and I was like 'Okay.' And I thought about it. I was like 'No. I wasn't in the wrong. I'm going to appeal this warning,' but there was no way that I could appeal."

The worst expression of algorithmic cruelty is disenfranchisement. Under the guise of safety, systems designers make it easy to block or remove an account in case a bad actor tries to cheat the system. This adversarial stance means that good workers are sometimes misinterpreted as shady players. Inevitably, mistakes are made. A worker changes an address, loses her internet connection, or shares an IP address with another worker. Each one of these things is a potential red flag. The algorithmic system sees the flag as a possible security threat and, with no one at the helm to distinguish friend from foe, the worker is penalized. The penalty may look like being blocked or suspended, or having an account deactivated. Again, in an ecosystem in which workers are seen as interchangeable, the system automatically eliminates what it deems bad apples. The sad irony is that even the best-intentioned and most seasoned workers can get caught in the dragnet.

Riyaz, 33, carries roughly 200 pounds on his 5-foot-10-inch frame. He lives a mile outside Vijayawada, a city on the banks of the Krishna River, in India's coastal state of Andhra Pradesh. On the day we met, in July 2013, the temperature was boiling hot, yet Riyaz wore a meticulously pressed blue-and-white button-down shirt, tan chinos, and black loafers. Still, his eyes looked tired and puffy. He held himself with the awkwardness of someone uncomfortable with the weight he was carrying. He mentioned that he wasn't sleeping well because of the stress of long hours, with little relief from looking for work. Members of our research team had taken a cab to his hometown from Vijayawada's one-gate airport—a drive that took about an hour. Per his directions, we met him at a corner store near his home. As we piled back into the cab, Riyaz climbed back onto his motorcycle, and soon our driver was following him down dirt roads, navigating around potholes, goats, and children until we arrived at a simple home near the far edge of the village.

Riyaz hadn't always lived in this small town. He moved to Hyderabad in his early twenties to learn computer skills and basic software engineering, with the hope that getting credentialed would give him entry to the city's booming IT sector, one of the largest in the country. As discussed in chapter 2, Hyderabad was one of the first cities in India to develop infrastructure to attract multinational companies engaged in software development and business process outsourcing (BPO). The city's well-educated and expanding middle class made it a natural location, as did the fact that Hyderabad's residents spoke English as readily as Hindi.

It was in Hyderabad that Riyaz first heard about MTurk. In need of cash, he signed up for the crowdsourcing platform. He did a range of tasks, from surveys posted by researchers to image tagging. He spent more than five months learning how to work on MTurk. One of the ways he found to succeed was by building relationships with reliable requesters. For example, Riyaz made a YouTube training video for one of his most frequent requesters. The requester liked Riyaz's work and wanted him to help train other on-demand workers in how to do search engine optimization. Riyaz, who'd always been one to go above and beyond for requesters, did the video for free, because he considered it an investment in his future.

Soon Riyaz was making more on MTurk than from any IT job he could find, being a young Muslim man with no contacts in the city. He was earning roughly $40 a day and identifying more jobs than he could handle. Back in his home village, he had a lot of friends and family members who had basic computer and English literacy but didn't have jobs. So he decided to move home and hire them to work on MTurk out of his family's home. Eventually the ten-member group would call themselves "Team Genius."

Team Genius thrived for more than two years. Then, in March 2014, Riyaz heard reports about some India-based workers having had their accounts suspended. He began to worry about the fate of his account. He had multiple accounts set up through his workers. He knew this could be an issue, but he also knew Team Genius was doing good work, so it felt to him like an arbitrary restriction. But then, one by one, the accounts held by members of Team Genius began to be suspended. Riyaz began to scramble like mad to find more work on any platform he could. He worked around the clock to support those who had become dependent on him and his contacts.

Then the thing Riyaz feared most happened. He got the following email:

> I am sorry but your Amazon Mechanical Turk account was closed due to a violation of our Participation Agreement and cannot be reopened.
>
> Any funds that were remaining on the account are forfeited, and we will not be able to provide any additional insight or action.
>
> You may review the Participation Agreement/Conditions of Use at this URL:
>
> http://www.mturk.com/mturk/conditionsofuse
>
> Thank you for trying Amazon Mechanical Turk.
>
> Best regards,
>
> Laverne
>
> P.S. We value your feedback, please rate my response using the link below.
>
> Amazon Mechanical Turk
>
> Please note: this e-mail was sent from an address that cannot accept incoming e-mail. To contact us again, select the Contact Us link related to your inquiry below.

Workers: https://www.mturk.com/mturk/contactus
Requesters: https://requester.mturk.com/contactus

Riyaz was instantly locked out of his account and his pay, with no clues as to how to appeal the decision or how to retrieve the money he'd accrued over the past two months.

As our cab pulled up in front of Riyaz's family home that summer day, we couldn't help but wonder if this was why he looked so exhausted. He parked his motorcycle as we trundled out of the cab. As we were collecting the cab fare, Riyaz waved the cabdriver away, with assurances that he knew the man and "would take care of it later." Before we could protest, Riyaz was leading us into the home he shared with his wife, their two children, and his mother. Inside the simple dwelling was a cement floor and walls painted sky blue. Thin curtains divided the home's main room into a kitchen, bedrooms, and a workspace for Riyaz. Like so many other makeshift home offices, it held an office chair and a small desk. An 18-inch LCD monitor running Windows XP sat on the desk's surface, while a rusty CPU cabinet squatted beneath. A trip to the restroom resulted in an informal tour of the home's back rooms—two bedrooms with twin beds adorned with colorful quilts. Freestanding clothes racks in each room overflowed with a combination of yellow, orange, and red saris alongside black burkas. Minutes later, Riyaz's mother greeted us in Urdu and encouraged us to gather on cushions around the family's low table for lunch.

As we talked about the fate of Team Genius, Riyaz's mother served a lunch of biryani, fish stew, and a spicy chicken dish. He told us how he'd lost his account a few weeks before. He shook his head. It became clear that he felt personally responsible for the livelihoods of nearly two dozen friends and family members. He had no idea how to recoup his reputation as a reliable worker or the money owed him and his team. Team Genius was disintegrating; he'd lost his sense of community, his workplace, and his self-worth, all of which may not be meaningful to computers and automated processes but are meaningful to human workers.

Riyaz told us how he'd reached out to the one person he'd worked with the most on MTurk, the requester for whom Riyaz had made a training video for other MTurk workers. In an email, he implored the requester to intervene. He asked the requester if he would be willing to contact Amazon on his behalf, to explain that he did good work, that he was a reliable worker, and that he deserved to keep his account. Riyaz never heard back from the man. When he told us the story, it was clear he felt betrayed by someone he considered a professional colleague. Riyaz even went so far as to write a letter to Jeff Bezos. That was a huge risk for Riyaz. The subtext of this work is that workers remain invisible. Standing up and calling attention to yourself is a fast track to getting a "bad reputation." Jeff Bezos never wrote back.

Unfortunately, Riyaz's case is not unusual. According to a national survey we conducted in partnership with Pew Research, 30 percent of on-demand gig workers reported not getting paid for work they performed. Workers can lose their job and wages, with no explanation and no opportunity to appeal the cancellation of their account. Companies decide how, or if, workers will receive final payment for work completed. People do not just lose their account access — many workers, like Riyaz, lose their livelihoods.

Is Riyaz's experience of losing his job through an "at-will" contract just the well-worn tale of what freelancers — the 1099 workforce — face every day? Or is it the harbinger of a new set of challenges and choices that we must all face — and redefine — if we care about the future of work for our generation and the generations to come? Should we apply the tried-and-true wages and hourly employment laws, or do we need a new set of rules?

Any freelancer will tell you — and the literature on independent contract work confirms this — that getting paid can be the hardest part of the job. In 2015, the Freelancers Union, in the United States, found that 70 percent of those freelancing in the current economy do not get paid by at least one client and 71 percent have struggled to collect payment for work at least once in the course of their career.[9] But the convoluted terms of employment for ghost work have made collecting one's wages even harder. Most freelancers and contractors have a point of contact at the company, someone to call or email if an invoice goes unpaid. They may even have a contact who will advocate on their behalf if a payment is late.

Conversely, on-demand workers must grapple with faceless platforms. No office manager, staff directory, or help desk offers assistance when things break down. On-demand workers operate as links in a long supply chain of labor, picking up pieces of a larger project and adding their polish before handing them in to be assembled later. Most people who represent links in the transaction will never meet one another.[10] So when a worker like Riyaz has an issue getting paid, he has no clear way to seek redress.

To complicate the scenario further, companies can also get into a bind. For instance, when an account holder or employee's identity comes into question, does a requester or company go ahead and place wages in a potentially fraudulent account or hold on to money owed, until the confusion can be cleared up and the rightful owner is identified? Even though the companies must wallow in a logistics quagmire, they are certainly in a better position to weather the uncertainty than the workers.

## No Boss Is Perfect. Neither Is Code.

Workers endure transaction costs, from finding a work platform to finding good work, all the way through to getting the work done while bearing the risk of not getting paid or having their account shut down for opaque reasons.

The vast majority of platform designers and requesters do not intend to be cruel. Platform designers are trying to deliver a seamless process to as many users as possible, including workers. No one is perfect, and neither is code. Computational processes often have unintended consequences. Programs are, as mathematician Cathy O'Neil argues, formally modeled "opinions embedded in mathematics."[11] They have no insight into what exceptions to the rule individuals may want or need. And requesters aren't obtuse out of malice. Like workers, many are learning as they go. They can be clumsy at using the site or poor at clearly communicating what they need. And they, too, are often under tight deadlines and demands from their bosses. No matter, as those on-demand workers were left to the whims of algorithms, which can be a cruel arbitrator as of late.

While requesters do endure some transaction costs, workers bear the brunt of them, with far more severe consequences, because they have so little power in this market compared with the requesters and the platforms. Workers have to be constantly on call to find good work, making them appear to requesters to be at their beck and call. Furthermore, these markets are extremely concentrated; for example, on MTurk, approximately 98 to 99 percent of all tasks are posted by 10 percent of the requesters, which exacerbates an economic power imbalance called monopsony.[12] In addition, many ghost work APIs are designed such that requesters post the wage for a task and workers have to either accept the wage or find a different task. There is no room for bargaining. All of these aspects of ghost work place more market power in the hands of the requesters.[13]

Because platforms get their revenue from the requesters, it is not surprising that they, intentionally or not, confer more market power to requesters. Platforms also have the power to unilaterally decide who does and doesn't have access to their platform. If a platform decides to use an automated process to freeze accounts deemed in violation of the site's terms of service, workers have no recourse.[14] Bad design is behind some of the dismissive, alienating, cruel-seeming functionality of platforms, but not always. In many instances, like providing more guidance or more direct channels of communication and training, platforms are not experiencing technical difficulties at all. As described in chapter 2, traditional employment used to take a long view of workers. Firms used to invest in individuals early in their careers to retain a steady, on-site workforce. Keeping costs down meant gaining loyalty and longevity, striving for workforce diversity to get the most of distinct perspectives; this model no longer fits today's highly specialized, always updating service and information economies.

If on-demand work is a harbinger of the future, it's noteworthy that traditional employment contracts are being replaced by a platform's "terms of service." The obligations in these agreements spell out the limits of what a worker can expect from the platform. They rarely detail how a worker would ever challenge work conditions, beyond deleting their account. And the absence of a physical work site makes matters worse. It's harder to document, let alone witness, how APIs exploit or exacerbate gaps in labor laws meant to protect workers' interests and rights. On-demand workers in the U.S. and India toil outside of any clear employment status. They have little access to protections associated with formal employment, aside from trying to make the case that they should be classified as full-time employees. And on-demand workers have nowhere to turn for advancement opportunities, accommodations for different abilities, laws to prevent discriminatory hiring, or legal recourse against unfair treatment, from wage theft to whistleblower protection.

Eventually all technologies break down. At some point workers need to connect with a human being to seek redress. Resolving that glitch —when an automated process needs a human not just to intervene but to care—lies at the core of getting ghost work to serve customers and workers alike. Since today's workers endure the brunt of the on-demand economy's transaction costs, weathering its most severe consequences, we focus next on how and why workers invest more in their work beyond rationally reducing the costs of business transactions.[15] As the next chapter attests, workers care about creating value, controlling their time and destinies, and discovering better matches for their interests and talents as much as they care about the size of their paychecks.

# 4

# Working Hard for (More Than) the Money

## Sizing Up the Options

Given how hard it is to find one's footing, let alone paychecks, doing on-demand ghost work, why would someone choose to do it at all?

As anyone whose air conditioner conks out in the middle of a sweltering heat wave can attest, sometimes there is no "choice" when life requires taking on extra work to pay one's bills. The Federal Reserve Board's annual *Report on the Economic Well-Being of U.S. Households* found, in 2016, that 40 percent of people in the United States did not have the means to cover a $400 emergency expense without borrowing money or selling something.[1] Signing up for an account and working from a home computer was, for some workers, the quickest route to fast money when they needed it most. They felt they had no other options.

But wouldn't working at the local mall or a fast food restaurant chain be better than ghost work? It depends on how one defines "better." Stable, decent-paying service sector jobs are not as easy to come by as a reader might imagine. And for those people experimenting with or committed to ghost work, the decision to keep doing ghost work, once they made it past the immediate need for emergency cash, is complicated.

Once workers met basic needs, they stuck with ghost work for reasons that were about more than the money. Those people did ghost work because they felt that it offered an escape route, or at least temporary relief, from the pressures and hurdles they had come to associate with a more typical day job.

Day jobs come with a mix of constraints and cultural cachet. Flipping burgers as a teenager to save money for a first car earns a pat on the back for being "industrious." Committing to that same job later in life does not offer the same validation or legitimacy, even though the paycheck is necessary and hard-won. Culturally speaking, taking an unpaid or low-paying job to build one's portfolio of experiences or to get a toehold on a career ladder as a writer, designer, or coder is validated as "entrepreneurial." Such effort is deemed worth the risk of a small paycheck, a worthy investment in one's future, particularly if the glamour of high tech is associated with it.[2]

One of the challenges for those doing ghost work is that there's no agreement about the social status or baggage that comes with it. Is it a dead-end trap, no different from the piecework of the first decades of the last century, or a hip gig that gives someone the ultimate flexibility? Is it better or worse than a "regular" job? People who've decided to pursue ghost work presumably did so after weighing the costs and benefits. They decided, at least for the immediate future, that ghost work was a better option. Their decisions hinged on both what they valued more than money and on taking stock of what "regular jobs" look like in their lives. Their job prospects reflect the growing and sobering reality of what is available to working-age adults around the globe.

The majority of the workers we met, in both the United States and India, are unemployed or underemployed at a moment in history when the vast majority of global job growth, as noted in the introduction of this book, is in food services, retail, construction, home care, and other service sector employment. Even in more professional careers, companies routinely start all but the most senior executive employees out on contracts, with some possibility of staying on but with no promises—from either the employer or the worker. In this way, employers have removed the lower to middle rungs of the career ladder and replaced them with temp work.

If plan A is a dream — the kind of 20th-century, salaried career track that's become less widely available — and plan B is the more widely available option of entry-level contract work, most likely a service sector job with low wages, few perks, no career trajectory, and unpredictable hours, then ghost work's design presents plan C. The premise of ghost work is that it needs people immediately available, willing, and able to answer or evaluate something that a computational process can't address on its own. In the absence of set hours, work sites, and professional gatekeeping, ghost work operates more like a self-organizing, organic online community. People are, in principle, invited to come and go as they please, rather than commit to a structured form of employment that tightly controls schedules, projects, and co-workers. And the system banks on redundancy — if enough people take a swing at a problem, the majority will signal the best probable answer to move the algorithms forward. That means that a person can earn money contributing to a ghost work's ecosystem, whether they put in four hours a week or 40, from one week to the next.

In the wake of this "open call" design — or chaotic mayhem, depending on the platform and the workers' level of experience — people use ghost work to gain some relief from the familiar pressures that dog traditional employment. They nimbly fit paid ghost work into their lives instead of forcing their lives into a regular day job. And some have kept at it, turning it into their main source of income, because it gave them some semblance of control over their time, work environment, and what they took on and valued as "meaningful work" to them.

In many cases, the context of India or the United States made little difference in shaping workers' reasons for entering the ghost workforce. The desire to control one's destiny and to be a part of a contemporary professional working world is a global, middle-class aspiration, after all. As this chapter illustrates, people found myriad, sometime idiosyncratic, ways to make ghost work meaningful and materially useful to them. They leave us clues about how to convert the shadow economy of ghost work into legitimately valued on-demand employment.

## When Career Ladders Lost Their Rungs

The picture many people hold about what a traditional job looks like — in terms of stability and predictability — is a product of the latter half of the 20th century. As noted earlier, not until World War II did organized labor and political clout combine (at least in some parts of the United States) to create full-time employment, meaning jobs that came with not just a paycheck but also stable hours, pensions, healthcare, and workplace safety.

Those jobs led to the growth of a middle class that had hit its zenith in the United States by the late 1970s. In the decades that followed, the middle class was hollowed out by deindustrialization and outsourcing.[3] What was left behind was the burgeoning growth of service jobs. This new form of employment rose from the thousands of retail chains, fast food outlets, and chain big-box stores that filled, first, American malls and suburbs and, not long after, their global equivalents. But service jobs weren't designed to replace the stable salaries and lifelong careers anchored to Cold War–era full-time work. Without the will among the business class to split profits with service industry employees or the strength of organized labor to push for the same safety nets put in place for manufacturing, service work arrived with low pay, uncertain schedules, long commutes from affordable housing, and a new set of customer service demands.[4] Working with the public was now a part of the job, too.[5]

As sociologist Gina Neff argues, by the beginning of the dot-com bubble, in the early 1990s, a generation of college-educated young people, particularly white men, faced a crowded job market made even more competitive by the GI Bill, post–Jim Crow, and second-wave feminism. More qualified applicants vied for a slowly draining pool of professional opportunities. But, as Neff points out, the meaning of "success" was also changing. Now "making it" came to be defined not by how far someone climbed up the corporate ladder and stayed around to earn a top pension but by how much someone set their own hours, drew stock option offers, won competitive bids for their labor on contract, or landed all three.[6] By the late 1990s and early 2000s, college-educated white-collar workers were throwing themselves into "venture labor"—a new world of high-risk, high-reward job opportunity associated with the tech companies of Silicon Valley and the stock options they hand out. Here success meant either cashing out early or, at the very least, controlling three things: when one worked, with whom one worked, and what kind of work one took on.

Generation X and millennial workers entered a job market that no longer offered the security of full-time employment with benefits familiar to their blue- or white-collar working parents. At the same time, their slightly older peers, at least the successful ones, now defined success as controlling their workloads or no longer needing to work at all. Demand for some full-time employees rallied back; by the mid-2000s, few were offered the well-paying, stable jobs familiar to a Gen Xer's parents. And by then, those with the most education or financial means no longer wanted jobs that anchored them to a timesheet or projects that someone else chose for them.

For those people with less education and wealth, having barely survived the global Great Recession of 2008, ghost work was a lifeline. They and their families worked multiple jobs, making distinctions between ghost work as a "primary" or "secondary" occupation meaningless. It was another source of income that they pulled together to cover their needs. Ghost work allowed them to escape from the inevitability of full-time service work.

## WHY PICK GHOST WORK OVER PLAN B?

Why are a growing number of people picking up on-demand gig work online instead of searching for some plan B that will pay the bills?

Let's start by stating the obvious reason for this. Despite reports of job growth, many workers can't turn entry-level plan B service jobs into viable stand-alone options. As noted earlier, many full-time service sector jobs don't offer wages, schedules, and locations that make them a "better choice" over ghost work. Consider the typical paycheck. Most jobs are mired in wage stagnation. Factoring in inflation, real wages in the United States were only 10 percent more in 2017 than they were in 1973, putting annual wage growth at a glacial pace of 0.2 percent a year over the past 40 years.[7] And the only reason that number isn't lower is that it's skewed by the wage spike among the very wealthiest wage earners, like the CEOs and Wall Street financiers. The top 1 percent of wage earners have seen their annual pay increase 138 percent since between 1979 and 2013, while the bottom 90 percent of workers saw only a 15 percent increase in their annual pay over the same period.[8] This means that the typical full-time job can't offer enough to be the sole source of income.

According to a 2018 report from the National Low Income Housing Coalition, there is no state, metropolitan area, or county where a worker earning that state's prevailing state minimum wage or federal minimum wage can afford to rent a two-bedroom home for their family working a standard 40-hour week.[9] A worker in Alabama, one of the states with the lowest cost of living, would need to make $14.65 an hour to rent a two-bedroom apartment. The state's minimum wage is half that: $7.25. At that wage, a worker would have to spend 81 hours a week to cover their rent. In fact, ironically, the typical service sector "plan B" job doesn't just pay too little — it also demands too much, namely, control over most of a worker's waking hours.

One in six people employed full-time have to contend with irregular work schedules. Ten percent of workers employed full- or part-time get their work schedules less than a week in advance.[10] Employees are sent their work shift updates via email, text, or phone call. They must be available to work the hours offered them. If an employee cannot rearrange childcare, skip other commitments like school courses or hours at a second job, they are paid only for the hours they can work and not given a chance to make up the hours elsewhere in the schedule. Despite loud criticisms in 2015 that led Starbucks, Disney, and other large companies to drop this practice, the trend, called "just-in-time" scheduling, remains common among larger retail and hospitality employers.[11]

As discussed in the first chapter, those more experienced at ghost work, once they've oriented to their platforms of choice and mastered their work routines, can earn hourly wages comparable to those of plan B traditional employment. It also allows them to avoid just-in-time scheduling or commutes to service-entry jobs that would compete with other constraints on their time.

So what, other than financial compensation, drives on-demand workers to take on on-demand jobs? Many of us might respond that we keep our day job, whether we like it or not, because we must earn money to pay the bills. Not surprisingly, most on-demand workers reported that "earning money" was the main reason that they work, too.[12] But how important is that money to them?

Throughout this chapter, we will use our interview data to show workers' motivations, which we augment with findings from our 1,729 surveys of workers across the four different platforms we studied. Combining these two data sets allows us to see how individual experiences "scale up" to the population level, suggesting something endemic to ghost work. Among the four platforms we studied, between 46 and 71 percent of the workers listed earning money as their primary motivation for doing ghost work. On the other hand, between 29 and 54 percent of workers said their primary motivation was self-improvement, such as gaining experience or learning new skills, or reasons of self-determination, such as utilizing their free time or being their own boss. While earning money is important, it's not the only reason workers do ghost work.[13]

According to a 2016 Pew Research Survey, roughly one-quarter of those doing on-demand ghost work reported that the money they earned was "essential" for meeting their basic needs. Another one-quarter said the money was "important." Of those who reported that the money was either essential or important, nearly half reported that they do this work because they have a "need to control their own schedule." Another quarter said there was a "lack of other jobs where they live." Those for whom the money was essential were also more likely to come from low-income households, more likely to be nonwhite, and more likely to have not attended college.[14] In fact, as existing research suggests, many workers don't have the financial breathing room to give up trying to make on-demand jobs work for them, because they have no other job options available to them.[15]

Yet even the most cash-strapped on-demand workers view their jobs as a choice, an employment decision that they consciously make for their own personal reasons.[16] Not surprisingly, when you ask people why they do this work, they'll say, "For the money." When pressed further, asked to share other reasons they stick with this work, which clearly offers less stability than a more traditional job, things get interesting.

Workers offered no single, dominating factor for turning to on-demand work. To the contrary, they valued on-demand work for a range of reasons, including the fact that they could pick and choose tasks and that they could stop working once they'd made the money they needed that week. It turns out it's unprecedented for large companies to organize employment this way.

## When Work Looks More Like a Book Club

Vilfredo Pareto was a famed 20th-century Italian scholar and a pioneer in the field of microeconomics. In measuring the concentrations and unequal distributions of income and housing access in social settings, Pareto observed that 20 percent of Italy's population owned 80 percent of the land.[17] Pareto's principle is a special case of the more general "power law" distribution used to describe the natural and social phenomenon by which a resource is concentrated in the hands of a few.

Pareto's formulation, the 80/20 rule, has been used to describe phenomena ranging from the distribution of income — the richest 20 percent of the world's population control roughly 80 percent of the world's income — to software engineering.[18] Microsoft engineers observed that fixing 20 percent of the bugs in a piece of software would take care of 80 percent of the glitches found in that computer program.

Pareto's 80/20 rule is also reflected in social systems. For instance, in a large book club, only a handful of avid readers show up at every meeting having read the book and feeling ready to share their thoughts on its broader meaning. Those are the 20 percent in Pareto's distribution. The remaining 80 percent of the book club contains two subgroups of members: those who show up having read some or most of the book but come primarily for the community, and those who show up wanting to try out the group. This last category of members may or may not stick around for the long haul, and they may or may not have read the book, but they are curious to see if the book club is a good fit for them. All three types of members are necessary for the longevity of a dynamic book club, but the 20 percent who are avid participants keep things steady.

This social dynamic of Pareto's 80/20 rule extends to online communities as well. Consider that most of the changes to Wikipedia are done by a small percentage of its editors. Or how, when you post an update to your Facebook newsfeed, a large percentage of your friends likely see the post but only a fraction will comment. The casualness of deciding when we opt in or opt out is what makes these communities work. Imagine if, in order to participate in Facebook, you had to comment on 100 percent of your friends' posts. Chances are a lot fewer people would be on Facebook.

When you think about choices and participation, it's only a small leap to see how the Pareto distribution applies to ghost work. But it helps to first see how it is incompatible with traditional full-time employment. With a traditional job, employers expect that workers will show up for the hours set for them and participate in the work fully for their entire shift. There is not a lot of choice. Workers know the terms and organize the rest of their lives around the hours they've been told to be at work. In exchange, companies give workers paychecks every two weeks. This traditional approach to labor subtracts the contingency — or the casualness — of worker participation.

But ghost work thwarts the traditional job structure. For the most part, there are no set hours, and projects are up for grabs and often assigned on a first-come, first-served basis. In this way, ghost work operates more like a self-organizing community — like a book club or Wikipedia or Facebook — and therefore adheres to a kind of Pareto distribution instead of a structured form of employment.

The Pareto distribution has long been a part of the labor market. Consider freelance writers, day laborers, and actors. A small, tenacious percentage are able to make a sustainable living; a larger percentage struggle to keep afloat, often with other jobs to shore up their effort; and the vast majority of people are testing the waters to see if it's a good fit.

What is unprecedented about ghost work is that large companies have come to rely on it to organize on-demand employment — a community of people with different investments, divergent interests, and diverse offerings, all treated the same and equally valuable to productivity. In the past, companies spent a lot of energy recruiting and retaining the best workers. For the first time, global companies are embracing — or at least unknowingly banking on — the Pareto distribution as a strategy for meeting their labor needs. Companies are populating their workforces by throwing open the front door, inviting everyone in, and then hoping some people stick around long enough to hit a project deadline, but no longer than that.

Ghost work platforms reflect a Pareto distribution in that a core group of the workers do the bulk of the work.[19] The specific percentages of distribution vary, in many ways depending on the platform's approach to retaining workers, but a Pareto distribution exists nonetheless. Ghost work currently organizes around a small percentage of people who turn project-driven tasks into full-time work. A slightly larger portion of people consistently contribute a few hours here and there as their schedules allow. And the majority of people come to the platforms to experiment and may find their way to intermittent or regular work, but they are just as likely to do one or two jobs and leave. All three approaches to ghost work contribute to the platform's bottom line. Even opening an account and having it added to the platform's "head count" — whether a worker is active or not—generates value for the platform, as it gives the appearance that the platform has a lot of labor on standby for those trying to find workers to fill a project need.[20]

For our purposes, we labeled these three groups *experimentalists, regulars,* and *always-on*. The vast majority of people start out as experimentalists.

Experimentalists are those who come to a platform but leave shortly thereafter, for a variety of reasons, including getting scammed or feeling exploited. Take, for example, Justin, whom we met in the introduction. He lasted no more than a month on MTurk. He learned about the site from his wife, who had friends in graduate school who'd used the site for their research. He describes MTurk as "exploitative" and beneficial only "to people living in places with poor economies." After a short stint, Justin did not go back to the platform.

Justin felt the tasks were exploitative, whereas other experimentalists simply found the platform too difficult to figure out on their own. These sites have a steep learning curve. New workers often struggle to decide which of the thousands of tasks available will be worth their while. Compounding the difficulty is that instructions for some tasks are hard to follow, and, as we detailed in previous chapters, unclear instructions can lead to work being rejected and the worker not getting paid.

"Regulars" are those doing ghost work who have found their footing but, for a range of reasons, work only intermittently. Some come back every few weeks, for a few hours a day. Others reliably come back, but only for a few hours a month of work. Importantly, all of them decide to stay in the mix, making themselves available, as time permits, without making ghost work their main form of income.

Regulars fit ghost work in between other facets of their lives and responsibilities. They might do this work between classes or around the edges of a full-time job. In some cases, workers did a few hours here and there at their full-time jobs, where they had reliable internet access.

Finally, "always-on" workers are those who turned ghost work into full-time jobs. These are people like Joan, in Houston, who lives with her elderly mother and needs to figure out how to make enough to pay bills and buy groceries. Typically, always-on workers need to control their schedules and have reasons, usually from past job experiences, to feel that ghost work is a better option than the jobs immediately available to them. One common theme we found through interviewing many always-on workers is the importance of having familial and social support to help navigate the complexities of on-demand work.

Just as with book clubs, wikis, and Facebook newsfeeds, all three classifications of participants are necessary to balance the ghost work ecosystem. The 20 percent of workers doing 80 percent of the work guarantee that the work gets done, and the remaining 80 percent of workers doing 20 percent of the work fill in the gaps.

## Fitting Work into Life Instead of Life into a Job

The arduous climb into the U.S. middle class, always hamstrung by one's gender, race, educational background, social status, and country of origin, is, statistically speaking, harder than at any point since the Great Depression. So it's not surprising that people turn to ghost work, despite all of its mercurial temperament, because they see something in it that alleviates the pressure of trying to jam their lives into more traditional forms of work, namely service sector jobs.

### STUCK AT THE OFFICE: BLOWING UP THE MYTH OF "WORK-LIFE" BALANCE

For some, ghost work is a means of escaping the confines of a cubicle. For others, most notably women, on-demand labor creates a side door into a respectable, legitimizing workplace.

Statistics show that once women earn their own income, a country's prosperity and health outcomes rise steadily. But, as gender studies scholars note, the past three decades of increased expectations that women *can* "have it all" — a place in formal employment, a family life, and personal well-being — have intensified demands that women *must* do it all, and without additional support. Gender parity in the home is rare and requires government support in the form of paid family leave and affordable childcare.[21] In studying mostly middle-class white women, Arlie Hochschild and, more recently, Melissa Gregg argue that achieving as a "career woman" becomes difficult when one is faced with the double bind of time constraints and expectations that accompany a full-time career and a full-time "second shift" managing and caring for a household.[22] Likewise, we found that women doing on-demand work in the United States and India struggled equally, though in different ways, to juggle demands on their time both at home and in the workplace.

There are striking similarities between the two countries. In India, with the growth of formal employment came a greater demand for women's presence in the service industries, particularly in business process outsourcing.[23] But the country's traditionalist impulses of political, religious, and caste ruling parties splinter cultural interest. India has made room to value the role of modern career women but is less accommodating of women who are financially independent and lay claim to identities beyond "wife," "mother," and "daughter."

While the pressure for women to "have it all" is a global phenomenon, women in the United States and India have different resources to navigate the constraints on their time.[24] Women are still expected to prioritize family and social obligations over their commitments to the job and may not be able to reap the benefits of a professional IT career, such as stable pay, family leave, insurance, and validation as a "career woman."

Asra is small and thin. She often keeps her eyes lowered until she gets to know you, but once she does, she is rarely without a smile. When she laughs, her hand covers her mouth, mirroring the modesty of her hijab. She dresses in full burka when out shopping or running errands in the crowded streets of the southern city of Hyderabad, home to one of India's largest Muslim Indian populations. She has two young children: a daughter and a son. Her husband is often called away to his medical clinic, serving their bustling Muslim-majority neighborhood.

We met at her house at the height of Ramadan — or Ramazan, as it is called in Urdu-speaking regions of India — for *iftar,* the sunset meal that breaks observing families' daily fast. She set the table with bowls of dates, large plastic bottles of water, and a soup tureen filled with freshly cooked haleem, a thick stew of spiced lamb, lentils, and wheat. We arrived early. We brought two boxes of cookies from Karachi's, a local bakery. She approached her front door, whispered "*As-salamu alaykum,*" and giggled as she added "Good afternoon." While writing in English came as naturally to her as writing in Urdu — both were first languages for Asra — she rarely heard English or had the chance to practice speaking it.

Asra studied engineering for four years in Victoria, Australia, where she received the equivalent of a master's degree. Then she returned to Hyderabad to marry Hassim. Now her priority is caring for her kids. Although she admits that chasing after children is boring, Asra is not interested in working outside of her home. In her upper-middle-class Hyderabadi Muslim family, she is expected to stay home. But, thanks to a brother who was interested in technology, Asra had been around computers most of her life. And her high scores in *Candy Crush Saga* and the word game Ruzzle reveal a competitive side. Knowing she loved computer games, her brother and husband encouraged her to work on MTurk. Her elder brother even helped her create an account on the platform, where she honed her ability to do image-tagging tasks for the requester called "Oscar Smith."[25]

When asked why she did MTurk tasks, she said, "Money." Asra paused, then chanted, "Money, money, money" to what sounded very much like the theme of *The Apprentice.* She watched movies and television shows in American English. Her house's generous-size rooms and the presence of her housekeeper, a Hindu woman in a bright-red, bangled sari, peeling potatoes into the kitchen sink, suggested that Asra's reasons for staying with ghost work were more complicated. Pressed about the need for money, Asra smiled and said, "If I earn my own money, I can buy gifts for my family. I can contribute. I am part of something. I'm able to work, like other people. But my office is my home."

Asra does ghost work for more than just the money. We turned to our survey data to understand how common her attitudes might be among her peers and the likelihood of U.S. and Indian workers doing ghost work primarily for the money. Our survey data shows that workers living outside the U.S. are more likely to do ghost work primarily for reasons besides earning money than are workers inside the U.S. One plausible explanation for this finding is that ghost work requires the up-front costs of a computer and an internet connection. Statistically speaking, if a person in India can access the necessary tools of on-demand work and has the requisite language and computer skills to participate in this online labor market, it's likely they already have some monetary resources and financial security before entering this workforce.[26]

For the past two years, Rajee, a Hindu woman, has worked roughly five hours a day on on-demand platforms. She lives in Coimbatore, in southern India, and stays up late at night to work on MTurk while her husband and two children sleep. She enjoys financially contributing to her household, too. Her productivity has improved her relationship with her husband, who, if he is awake while she's working, will bring tea and dote on her while she is at her keyboard.

More than the money, it's this familial acknowledgment of Rajee's contribution that means the most to her. She also enjoys being part of something bigger. She is active on closed Facebook groups for India-based workers and likes meeting fellow workers in the groups. Being a part of a community feels good, even when she's sitting alone at her laptop.

Asra and Rajee exhibit the temporal juggling act of many caregivers. Because it can be molded to fit into nearly any schedule, on-demand work offers them a way into the job market. Our survey data showed that

U.S. workers are more likely to work during the day than Indian workers. Since many of the companies that post jobs are based in the U.S., they usually post during U.S. business hours. Hence, workers in the U.S. can work during typical "nine-to-five" hours, and workers outside the U.S. have to adjust their scheduling to access more tasks.[27]

Our survey data corroborates Asra's and Rajee's stories. It shows that, while women and men do ghost work roughly the same number of days and hours per week, they differ in how they spend their time. Overall, men are more likely to do ghost work during the nights and weekends, and women are more likely to do it during the day and less on the weekends. If we assume that males are more likely to work a typical nine-to-five job outside the home during the day and during the week, and women are more likely to work inside the home during the day and during the week, a pattern begins to emerge. Women are more likely to do ghost work when their familial and household responsibilities permit. Men, on the other hand, are more likely to do ghost work in the evenings and on weekends, after they have fulfilled their outside-the-home work responsibilities.[28]

Lalitha, a Christian mother of two, also lives in Hyderabad. She left her call center job after getting married and later joined LeadGenius, seeing it as an opportunity to work without having to commit to a full-time position outside of her home. She enjoyed the work and did it well. But she turned down a promotion to junior manager — a post that requires workers to commit to both weekend and night shifts — because she did not want to compromise the care of her two children, what she considers her primary responsibility.

Lalitha illustrates something we saw again and again. Not all workers see on-demand work as a stepping-stone to advancement or even to steadier (eventually full-time) work. Instead, ghost work becomes a way to earn money and feel a sense of financial and personal independence in a life full of obligations to others. Our survey data showed that non-U.S. workers are less likely to report that the monetary reward is their top reason to do on-demand work. When earning money is not the top priority, workers do ghost work both for reasons of self-improvement, such as gaining experience or learning new skills, or for reasons of self-determination, such as utilizing free time or being one's own boss.[29]

There are two ways to interpret the gendered labor of on-demand work platforms and the value that women derive from this kind of work. The first view might celebrate on-demand jobs as opportunities to free women up to have it all, imagining that on-demand work is the answer to the working woman's dilemma of needing to leave the home to earn an income. The second view sees this work as perpetuating traditional expectations of women to handle both full-time family obligations and the workload of more formal employment.

Both views are equally valid reads of the situation. For some people we interviewed, particularly women, ghost work legitimized their contributions and gave them a way to feel valued. Women are not unaware that their options are limited in formal employment, at least in part, by the expectations that they will continue to manage their roles as wives, mothers, and adult children caring for aging parents. In the absence of other options, like fully subsidized parental leave and childcare, that make it easier for parents to equally share household responsibilities, women and men turn to ghost work as a means of getting out of the office as they strive to balance commitments to family and other dimensions of their lives.

## HUSTLE

Like entrepreneurs aspiring to greater fame and forgoing pay as part of that path or those hoping that their online media productions will pay off, some use on-demand ghost work to build an on-ramp to a new career.[30] Ghost work can become a stepping-stone or easy-to-access on-the-job training. On-demand work becomes a "sandbox" where people can practice things like graphic design, typing, transcription, computer literacy, and language translation. These experiences are more difficult to acquire in a more traditional workplace, where there are greater expectations and pressure to perform.

Virginia has a bachelor's degree in international studies and a master's in global affairs. As a native Spanish and English speaker, she wanted to put her language skills to work at the UN or another nongovernmental organization, advancing a peaceful mission through cultural exchange, but she struggled to find an entry-level job. When Virginia became a project manager with Amara two years ago, she had opportunities to develop her Spanish-English translation skills as well as to practice learning several more languages. "I can speak decent Arabic and French now! I couldn't do that with any other job. It's like being paid to go to language classes every day!" Virginia saw her work at Amara as a way to build her dream career. "I can take what I'm learning anywhere that I want to go next." She effectively uses Amara to create value and meaningful work for herself, and Amara's worker-centered focus allows her to transform ghost work into decent on-demand employment.

Like Virginia, Gowri, a 23-year-old living in the small town of Erode, in south-central India, saw MTurk as a chance to practice and to build skills she could use for future work. The oldest child of parents who teach and sell their weaving, Gowri decided to try on-demand work through a course at a local computer center that taught her how to set up an MTurk account. Her focus: to improve her English and basic computer skills. "I can write English, but it is hard to learn everyday English phrases through newspapers and magazines. MTurk tasks let me look terms up and practice searching for information, like postal addresses in other countries, that I would not otherwise know how to do at school."

For now, Gowri is focused on saving money for her upcoming marriage and learning the communication and computer skills that might give her access to better-paying work. "My typing has gotten so much faster doing this work. That is a skill that I can use if I go into accounting or take exams for finance. Really, everything that I'm doing now, I will do when I have a job in finance or at a bank."

These stories reflect the value that workers put on building their skills through doing on-demand ghost work so that they might be able to get a better job down the road. Gowri's desire to develop her typing and English skills shows that, while the overall trend might be for less-educated workers to primarily do on-demand work for the money, there are certainly exceptions to this general trend, and she is one of them.

Our survey data shows how the outside options workers may or may not have — due to the number of other income sources they have, their age, or their educational background — are associated with their motivations for doing on-demand work. First, workers who have more income sources besides ghost work are more likely to do ghost work for reasons beyond earning money. Second, workers who are younger are more likely to do ghost work primarily to gain experience or to learn new skills, as opposed to doing it simply for the money. Finally, workers who are more highly educated are more likely to do ghost work either for self-improvement, such as gaining experience or learning new skills, or for reasons of self-determination, such as utilizing free time or being one's own boss.[31] Taken as a whole, these results suggest that workers who have other options to earn — whether because they are younger, have more education, or have other income sources — are more likely to give something besides earning money as their top reason for doing on-demand work. For others, ghost work offers a way to support an interest that has yet to provide a steady income.

For example, Carmela, 30, relocated from Florida to Chicago to follow her dream of becoming a choreographer. In the past, she'd earned money by teaching dance and working as a brand ambassador, representing a company and marketing its products at events. Both part-time jobs paid the bills, but neither fed her dream. Teaching dance meant she had to adhere to a schedule and couldn't travel freely to pursue her choreography. Being a brand ambassador left her feeling empty; she calls it a dead-end job. "There's always going to be a company that needs their product pushed. It's not like it's going to lead to anything else. It's not benefiting me, as far as my career."

Like Virginia, Carmela had grown up speaking Spanish and decided to take advantage of the fact that she was bilingual and loved languages. She enrolled at a community college where she took classes in language interpretation and translation. Then she decided to volunteer to translate and subtitle TED Talks for practice as part of the Open Translation Project. From there she discovered Amara, just as the company started offering paid translation and captioning work on demand. Even though Carmela can earn more money as a brand ambassador at a corporate event in less time than it takes her to finish a transcription project for Amara, she chooses to work on Amara. "I can make money wherever and work on things that matter to me." She adds, "I'm not really looking to springboard into anything else. I want to be able to travel to pursue choreography. All I need to do is take my computer with me, and if I get a job assignment I'm still working. I'm living my ideal life."

Formal employment in the service sector—the kinds of work most readily available to both Carmela and Virginia—tie paychecks to weighty obligations. They tether people to specific physical locations in exchange for decent-paying work. The long hours or emotionally empty work can drain energy from projects, paid and unpaid, that they enjoy. Workers can make ghost work a navigable path out of challenging circumstances, meeting a basic need for autonomy and independence that is necessary for pursuing other interests, bigger than money.[32]

## GLASS CEILINGS

On-demand jobs offer those in the U.S. and India who face workplace discrimination—particularly historically marginalized communities, women, and people with disabilities—digital literacy, a sense of identity, respect among family, and financial independence. Women who dropped out of the workforce to care for young children face barriers when they try to return. Women in the U.S. and India come from different religious and socioeconomic backgrounds, educational levels, and social roles, but women in the two countries share similar challenges in receiving fair pay and recognition for their contributions in the workplace, at the same time that they, paradoxically, go unpaid for their irreplaceable work as caregivers in their households.[33]

Kumuda, 34, is a Hindu mother of two who lives in Chennai, a coastal city in Tamil Nadu. She has a high school diploma in electronics—no small accomplishment, given that she was born into a lower Hindu caste in which women often do domestic work for higher-caste families. She says she owes her education to her father, who kept her and her sister in school long past the point at which most girls in her village are kept home to work rather than continue their schooling. Her father's decision didn't go unnoticed by others in the village, and he was chastised for allowing his daughters to stay in school. The fear was that it could work against the family when it was time to arrange a marriage for Kumuda, as it's harder to find a match for an educated woman of her caste and class background.[34] But her father held strong.

Now Kumuda's diploma in electronics qualifies her to teach at a local computer training center. She also earns money teaching spoken Hindi. But her biggest source of income is ghost work. When Kumuda started working on MTurk three years ago, Kumuda's husband and in-laws were cynical. How could she make any significant money sitting alone, shut away in a back corner of the house, hunched over a laptop, completing jobs issued from companies in the Pacific time zone? But after her income matched, then surpassed, her husband's earnings as a repairman, she gained the support of her extended family.

Kumuda's earnings—nearly 25,000 rupees (roughly $350) per month—make her the highest earner in her town. Her dream is to earn enough money to start a coaching center, named after her father, so that all the members of her village will see the value in educating young women. "My father wanted more for me than he had growing up. Seeing me succeed—become the highest earner in my village—made him very proud."

Danelle, 35, spent a few years finishing her coursework and exams toward earning her doctorate in biochemistry. She experienced sexism in her traditional scientific work environment. That, plus the demands of being a mother of two, made her decision to work on LeadGenius a more attractive option. She did ghost work for the company during its early days, when it was known as MobileWorks, and when the company got its second round of angel investing, it hired Danelle to be the office manager. She happily moved her family to Berkeley, California, to work as a full-time employee in the company's main offices. She describes LeadGenius as an incredible and inclusive workplace.

Despite being a world apart, Danelle and Kumuda both show how on-demand labor can have a transformative effect not only on workers themselves but also on their families. But it's not just women like Kumuda who face glass ceilings. People who had faced discrimination in the workplace because of disability, sexual orientation, or gender identity reported that on-demand work was a way to avoid harassment from co-workers with more seniority or power over them.

Lakshya, 34, was in an auto rickshaw accident years ago that left him paralyzed from the waist down. He lives with his immediate and extended family in a lavishly furnished home, in a well-established East Delhi neighborhood. Before the accident, he was a mechanical engineer, sending much of his income to his parents to help them buy the land and build the house that he lives in now. He spends most of his time upstairs, in a large corner bedroom with a balcony that looks out over the house's gated entry. Family members carry him about the house, up and down the stairs, but he rarely goes out.

After recovering from the accident, he looked for work for at least a year, but, after so much rejection, he turned to online work. At least no one online would see his disability. Although it has been illegal to discriminate against people with disabilities in India since the 1990s, it is not uncommon for people like Lakshya to feel pushed out of formal employment, as happens in the United States. He can never know if he's been passed over for a position because of his disability or because of the gap on his résumé from the time he spent recovering in the hospital, which totaled more than a year.[35]

For nearly two years, Lakshya has been doing ghost work on UHRS.

He does more than 150 tasks an hour on average, and in the previous month he had worked almost 200 hours on the site. Lakshya works on categorizing news stories, reviewing adult content, categorizing video content, categorizing the words that people use to search for items on Bing, and voice comparisons of British and Indian English. He also trains chatbots — voice- or text-driven computer programs used as intelligent conversational agents — to recognize differences between someone asking a question and someone making a statement, completes short marketing surveys and image relevance tasks, converts search questions into conversational forms, helps improve queries in Hindi, and reviews adult content captioned in Hindi. "I do it to keep my mind active," he says. "I have to do this. I have to keep busy." There is urgency in his voice.

People like Lakshya and Kumuda use on-demand jobs to pry open employment opportunities that might otherwise shut them out. Platforms like UHRS and MTurk provide very little, if any, data about the workers to the requesters. Recall from chapter 1 that the API abstracts away a worker's individual characteristics. All the requester knows is that worker ID A16HE9ETNPNONN did the work. The requester doesn't know if worker A16HE9ETNPNONN identifies as a man or woman; Muslim, Hindu, or Christian; disabled or not. The downside of this abstraction is that it's dehumanizing and can make requesters forget they are even hiring humans. The upside is that requesters can't as easily discriminate against Kumuda because she's a woman, or Lakshya because he is paralyzed.

When we started our research, Hindu nationalism was on the rise in India. There was an historic shift in party power from an ostensibly leftist, secular Indian National Congress to the Bharatiya Janata Party (BJP). Through the eyes of our research participants, we saw how this shift punctuated the explicit demands on women's allegiance to family, religious, and cultural obligations. Perhaps the political conservatism of the moment made on-demand ghost work more meaningful to the Indian women we met. Ghost work became a conduit to the oft-sought-after role of modern working woman. They could take on earning an income without throwing themselves into the tense national debates surrounding the impropriety of "call center girls" — women who work swing and night shifts, alongside men of all castes and religions, and who are often accused of prioritizing making money over propriety and piety.[36] At the same time, women in the United States were just as likely to talk about the value of on-demand work as a way to control their economic destinies, break into new work, or build up new skills while balancing childcare and eldercare.

People use ghost work to counter familiar pressures that come with making a living. But, turning to ghost work as an alternative has its limits, particularly for those experimentalists unable to find the resources or peer support to learn how to find a pace that works for them.

## It's Not All Sunshine and Roses

On-demand work's Pareto distribution offers people a chance to tailor work around life commitments and get some relief from the pressures of a more typical plan B service sector job. As the workers' stories above attest, on-demand work is not inherently a bad gig. It can be transformed into something more substantive and fulfilling, when the right mixture of workers' needs and market demands are properly aligned and matched. It can rapidly transmogrify into ghost work when left unchecked or hidden behind software rather than recognized as a rapidly growing world of global employment. Technologies, in and of themselves, are not great equalizers. The most obvious check on the potential of on-demand work as an economic opportunity for everyone is that half the global population doesn't have access to it. On-demand work cuts anyone without a reliable internet connection out of the picture. If countries keep pace with current rates of internet growth, 100 percent global internet adoption is still two decades away.[37] And among those connected, the majority of the planet still accesses the internet on a woefully slow connection and uses outdated devices.[38] Despite relatively affordable 4G and broadband rates in India, many workers we interviewed struggled to maintain their work hours during monsoon season, when heavy rains and winds frequently interrupt power grids. The world's working adults are not outfitted for ghost work, and no specific employer or government agency is tasked with changing that fact.

The other challenge is that, with so much variation in workers' schedules and commitments, the "flexibility" of ghost work's Pareto distribution means that people don't share a work site, hours, or a professional identity—three key ingredients to organizing workers' interests. Lastly, the lack of coordination among workers to stabilize the price they put on their labor, combined with the power requesters have to price work so that the lowest-bidding worker "wins," means that requesters can often find a pool of workers willing to do things more cheaply than others, which drives wages down for all workers on the platform.[39] That people weather the downsides of ghost work says more about the shortcomings of plan B employment than it does about the upsides of ghost work. It also reminds us that people will always find ways to make their work meaningful.

We found that workers doing on-demand ghost work, like workers everywhere, have more in mind than getting paid when they take on a job. Caring about something other than a payday is a way to feel some measure of power, control, and autonomy in a world where economic pressures curtail fully chasing our dreams with reckless abandon. One of the most common ways people make the economic necessity of work meaningful is through social connections and camaraderie.

# PART III

## Talking Back to Robots

5

# The Kindness of Strangers and the Power of Collaboration

## Introduction

Anyone who has ever started a new job knows how important it is to ingratiate yourself to your co-workers. These are the people who show you the bathroom, explain how to fill out a time card, and offer advice on how to handle a delicate conversation with the boss. Eventually, your co-workers may even share information about other jobs, such as an opening in another department, or a manager with a reputation for mistreating his workers, someone you'll want to avoid. These watercooler conversations extend the social space and intimacies of the office in ways that are highly valued but easily taken for granted.

With that in mind, consider how it feels to be an on-demand laborer. Your office may be your kitchen countertop or a wobbly table pushed against a bedroom wall. Most likely you toggle between working and caretaking for children or elderly family members. Depending on where you live, you may or may not have reliable electricity or internet access. If cellular service is cheaper than broadband, your phone may be your only portal to work. Your point of contact if you are an employer is not a person but a computer interface. Most jobs pay only a few cents, but you hope that if you do enough, the pennies will add up. When you complete a job, you never hear "Nice work!" so you must be your own cheerleader and taskmaster. Well-paying tasks, those that creep toward the equivalent of $15 an hour, are fleeting, but you scan the job boards day and night in hopes of catching one.

By design, ghost work attempts to strip a job down to its bare necessities: an assignment and a payday. Designers of on-demand labor platforms assume "users" work independently and autonomously. To them, workers are one piece of the bigger puzzle of how to offer goods and services quickly and efficiently to consumers. Digital labor is a means of collecting data to feed into an algorithm or producing content that is good enough, fast enough to meet an urgent deadline. This view isn't nefarious or malicious, but it's blind to the reality and demands of ghost work.

We looked behind the grand facade of on-demand labor platforms by circumventing companies' APIs and meeting workers face-to-face. They told us about how they built their own complex and thriving social network to combat the isolation of doing on-demand work. We saw workers trading information online, calling one another when well-paying gigs posted, and sharing computers and internet connections when someone was just getting started. We heard stories about workers collecting money to replace a colleague's broken computer. We heard about gifts exchanged and birthdays celebrated by people who'd never met one another face-to-face. Workers collaborated both to devise work-arounds to the platforms' technological shortcomings and to address their social needs as people sharing the slog of a workday.

The need for social ties associated with traditional nine-to-five employment persists, even in the absence of a physical workplace. Because there is no infrastructure for workers, they create it for themselves. Workers collaborate to overcome social and technical problems stemming from the platforms they work on. More specifically, they collaborate to reduce the transaction costs imposed on them by the platforms to get the work done and provide social support to one another. The most engaged workers invest in three types of collaboration. First, they collaborate to reduce overhead costs, such as signing up for an account, avoiding scams, finding work, and getting paid. Second, they collaborate to get the work done. And, finally, they collaborate to re-create the social side of work. Ultimately, to be good at their jobs, workers desire, and arguably depend on, the social aspects of work — the gab, the gossip, the informal trading of tips and support. Ghost work relies on the kindness of strangers.

Workers use collaboration to puzzle through, pragmatically, what to do in a setting that has little guidance.[1] They also take comfort in the relief of others sharing the cognitive load of feeling lost, much like talking through a work frustration is, practically speaking, a way of moving work forward.[2] Collaboration among on-demand workers lays bare the deeply social experiences and meanings that work produces in our lives.[3] Workers' penchant for collaboration also highlights how the presence and effort of people behind the scenes are the true currency and value of this labor market. Contrary to our stories about scale happening through the power of fine-tuned computer code, workers build off APIs and one another's efforts. Throughout this chapter we'll return to the importance of collaboration and how it relates to how work gets done in settings that put little stock in the need for workers to talk with anyone at all.

Collaboration happens most among those workers spending the greatest number of hours doing this demanding work. Although it's a cross-cultural phenomenon, collaboration among on-demand workers in India and their American peers looked different. The data in this chapter reflect that reality. We saw that in India the need for texting and physically sharing the same space echoed other hard realities, like unreliable electricity during monsoon seasons, as well as the linguistic challenges of translating colloquial American English phrases like "jonesing for a cookie."

While this chapter will give greater attention to the ways in which Indian workers illustrate the importance of collaboration, we note how workers in the U.S., in their own ways, are as likely as their Indian peers to collaborate. And we'll end the chapter with a poignant example of a time when workers — in a collective show of strength — collaborated across oceans to advocate for themselves and one another.

## Collaborating to Reduce Transaction Costs

Joseph, 22, is a Christian student in the south Indian city of Thiruvananthapuram. He has a bachelor's degree in computer applications and wants to pursue a master's in design, but his real passion is music. Joseph plays the guitar in a local band. When he isn't practicing or performing, he works on MTurk. He joined MTurk in 2012 because he could earn money and still have plenty of time for his music. He'd tried other online job sites, but they turned out to be fake. He heard about MTurk through a Facebook group. His first request to open an account on MTurk was rejected, because his identification papers listed an old home address. Here's where it's helpful to remember that in order to sign up for MTurk, workers must have government-issued identification that shows a home address.

What seems straightforward to many Westerners is often a hurdle for workers in India, where people's homes often lack a formal address. This was the case for Joseph. Unable to offer a postal address that satisfied MTurk's requirements, he bought an account from an agency and used its address — a common work-around. Shortly after Joseph bought that account, it, too, got suspended, likely because the address didn't match his government-issued identification. The third time Joseph tried to get an account, he bought one on Facebook. As he put it, "I found a person on Facebook who was from Thrissur [a city two hours away], but he does not work on it [MTurk] anymore. I work using his account and give him 20 percent of my salary."

For the past four years, this arrangement allowed Joseph to work on MTurk and keep 80 percent of his earnings, roughly 20,000 rupees ($275) a month. The person who sold Joseph the MTurk account received the paper checks, deposited them, and transferred 80 percent of the earnings into Joseph's bank account. Joseph used the money to contribute to his family. He supported his dad's stationery shop, and he bought his mother her first washing machine for her birthday. He also indulged in a motorcycle for himself. Without the cooperation of the account seller, whom he now considers a "work friend," Joseph would not have been able to manage the administrative requirements of having a validated physical address and access to a bank for processing his pay.

Joseph's story illustrates a common scenario in India. Opening an account on a digital labor platform is more complex than it might appear, especially for people in areas where there are no roads or government services to guarantee their residences will receive water, much less mail. At the time, Amazon paid on-demand workers with paper checks. Yet several of our participants described having trouble getting postal delivery, much less being able to deposit checks, since their neighborhoods didn't have banks. Like Joseph, some people we interviewed described using a friend's or family member's physical address because it was more reliable, but doing so put their account at risk of being suspended if the address didn't match their identification papers.

Kumuda, 34, found MTurk after falling for a job scam. Now she helps relatives and friends avoid the same mistakes by guiding them to trustworthy platforms. As she says, "I started with outsourcing. I and a friend of mine were searching for job offers. Everywhere it was a scam. I am the first person in my area to find out about MTurk. My friends have come to know about MTurk through me [so they know it is safe]." Kumuda is an example of a worker who collaborates to help others find work. Opening an account on a labor platform requires workers to hand over intimate financial information. In the absence of a physical place of employment or any clear system for vetting the authenticity of a platform, workers must rely on one another to discern legitimate businesses from scams looking to glean workers' emails and personal information.

We found that roughly 25 percent of workers in the U.S. and India were referred to MTurk by a friend. Workers on LeadGenius had even higher rates of referrals, suggesting that word-of-mouth employer recommendations cut across platforms.[4] Workers share tips about tasks via phone, forums, chat, Facebook, and even in person, especially if they pay well. Likewise, names of requesters (aka employers) who are fair and reliable often circulate among on-demand workers.

Sanjeev, 22, is a student in Kerala who makes money by working on MTurk and maintaining several blogs on love and friendship that generate passive income through Google AdWords.[5] He first learned about MTurk from a friend in his computer applications course. He likes MTurk because he can do it part-time. He often keeps an eye out for late-night tasks, something he can do while he studies. When he sees a promising job pop up, he shares the news with his friend. He says, "If I'm working and find a good HIT [job], I call him and tell him about it."

While our ethnographic and survey data suggested that workers communicated and collaborated, we couldn't see how common or widespread these practices might be or how attributes like geographic location correlated with the amount of collaboration behind on-demand work. So, to understand the scale and structure of this effort to collaborate, we asked MTurk workers to help us map their *entire* communication network. Our HIT was a "Facebook-Lite" for MTurk workers in that it allowed them to anonymously report the online nickname of who they communicate with and the medium they use to communicate, such as email, SMS, Skype, or one of the many online forums workers use.[6] Workers could see a variety of information about their connections, such as why the worker started working on MTurk and how they stay motivated, along with whatever demographic information that worker felt comfortable sharing.

The network consists of 10,354 workers and 5,268 connections between them. There were 1,389 workers (13.4 percent) with at least one connection to another; among these, the average worker was linked to at least seven others (see figure 2). At first glance, one might wonder why we care about a network where 86.6 percent of workers are not even connected to anyone. The 13.4 percent who are connected are the hard-core workers, the always-ons and regulars from chapter 4, so when you put up a task on MTurk, you're more likely to get workers from this pool. Furthermore, we'll see that workers congregate in different online fo-

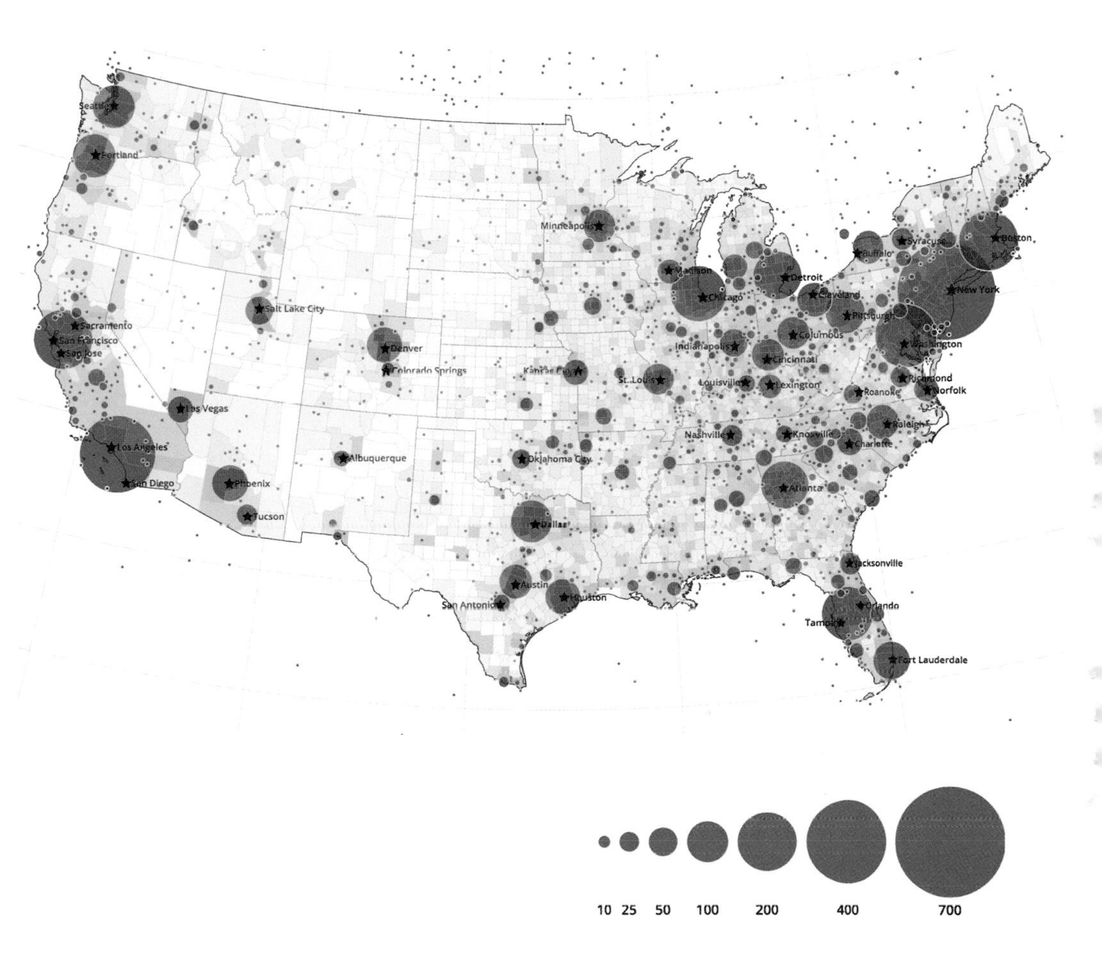


FIGURE 1A: Self-reported locations for participants on MTurk from March 14, 2014, through May 28, 2014. The number of workers reported in each area is proportional to the area of each circle. For example, the circle around Charlotte represents 53 workers, Atlanta represents 158 workers, and New York represents 671 workers. To protect the workers' privacy we randomly moved the locations they reported by a small amount. Coloration of counties indicates population density. *Gregory T. Minton*

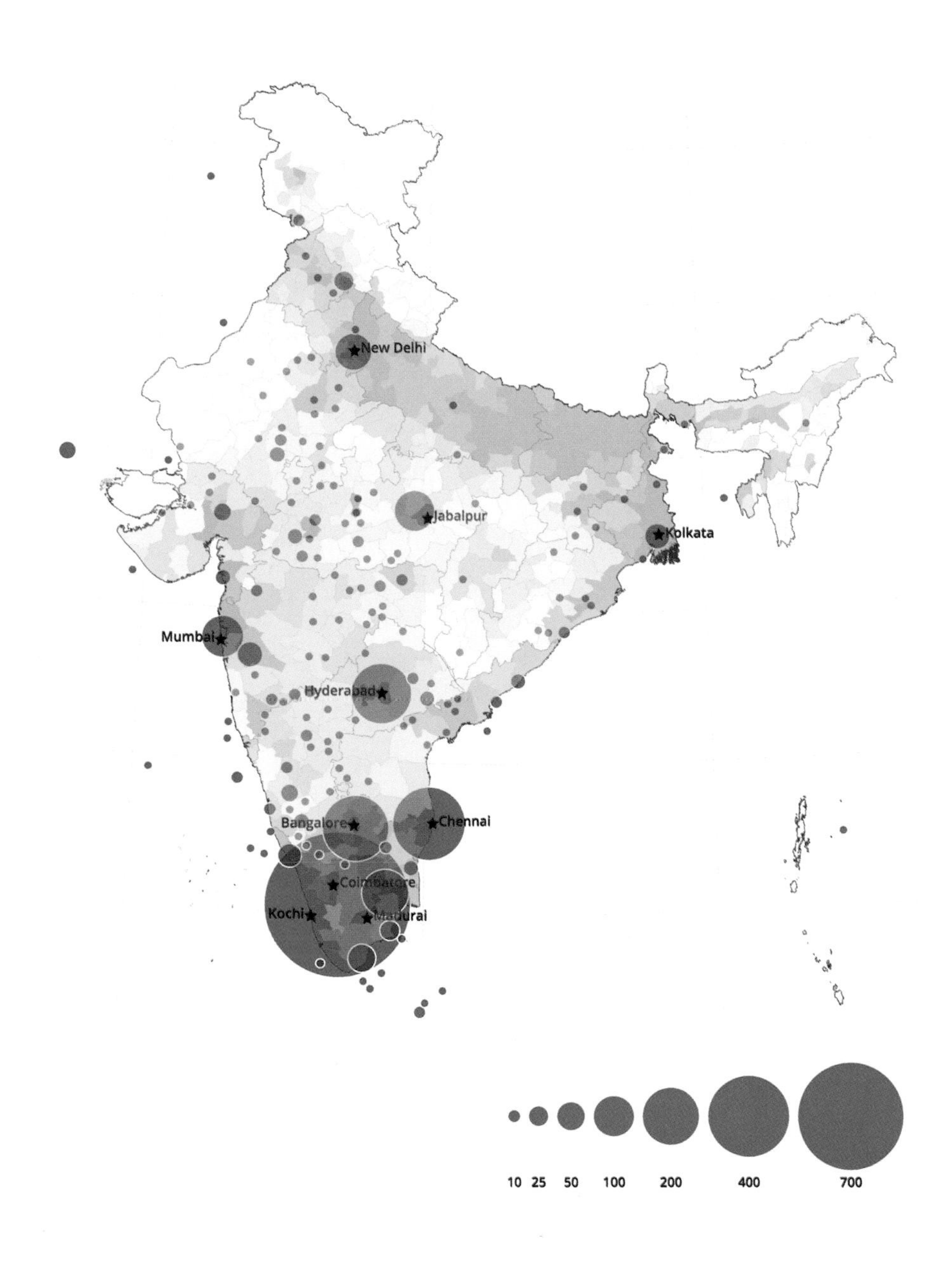


FIGURE 1B: Self-reported locations for participants on MTurk from March 14, 2014, through May 28, 2014. The number of workers reported in each area is proportional to the area of each circle. For example, the circle around New Delhi represents 17 workers, Hyderabad represents 46 workers, and Coimbatore represents 265 workers. To protect the workers' privacy we randomly moved the locations they reported by a small amount. Coloration of districts indicates population density. *Gregory T. Minton*

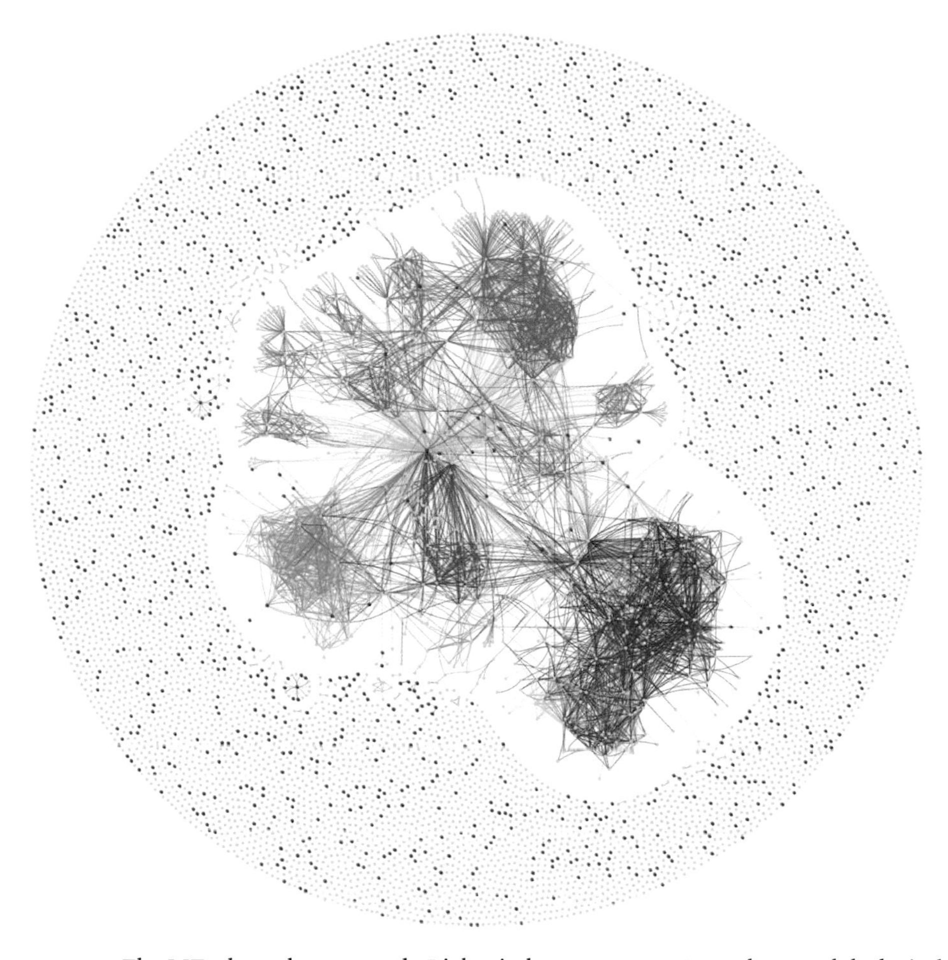

FIGURE 2: The MTurk worker network. Light circles represent U.S. workers, and dark circles represent non-U.S. workers. Light-gray links indicate communication via one-on-one media, such as phone calls, emails, text messages, instant messages, or video chatting. Colored links indicate communication via an online forum, where pink represents Reddit's HITs Worth Turking For, red represents MTurkgrind, orange represents TurkerNation, blue represents Facebook, and green represents MTurk Forum. *Ming Yin*

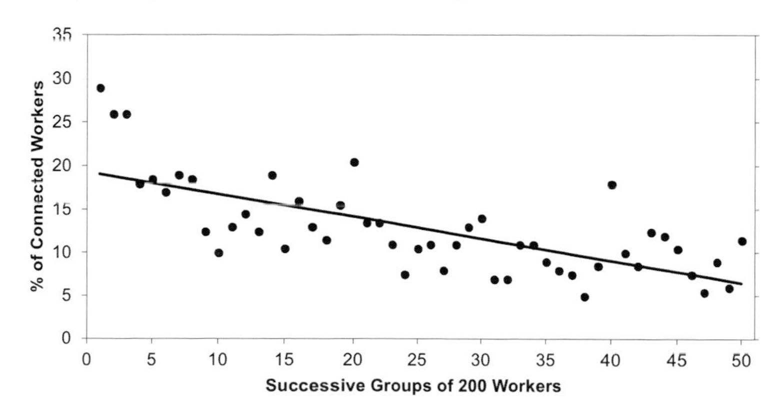


FIGURE 3: The percentage of connected workers who did our network mapping HIT (shown in figure 2) in successive groups of 200, ordered chronologically. *Ming Yin*

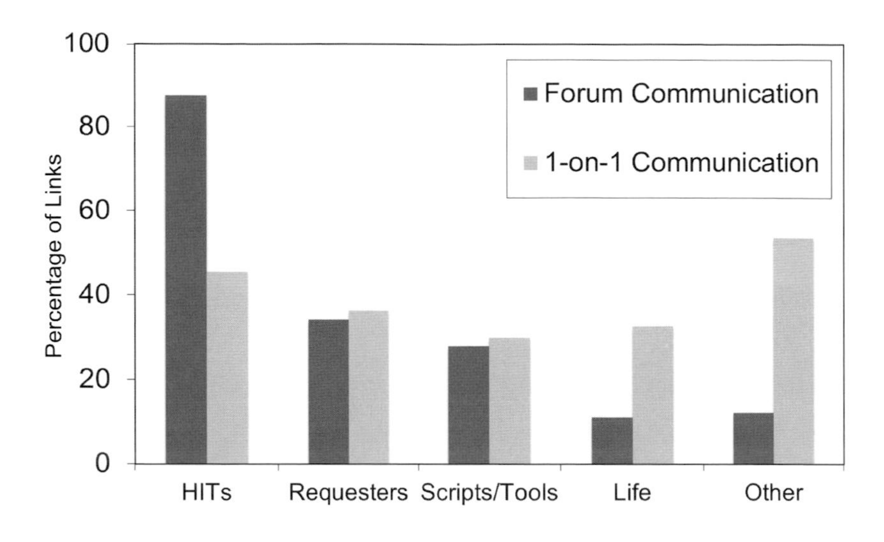


FIGURE 4: Comparison of topics discussed via online forums versus one-on-one channels. *Ming Yin*

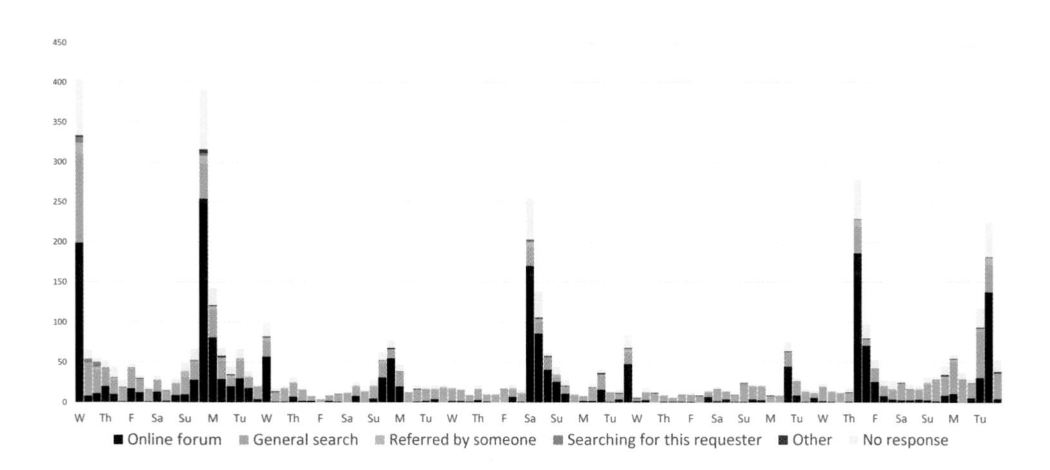


FIGURE 5: How 4,856 workers found our geographic mapping HIT (shown in figure 1) from April 23 to May 28, 2014. The x-axis shows each eight-hour period for the five-week duration. The y-axis shows the number of workers who did the HIT during that eight-hour period. The black region of each bar indicates how many workers came to our HIT via an online forum in that eight-hour period. The blue region of each bar indicates how many workers came to our HIT via searching the MTurk site in the same eight-hour period. *Gregory T. Minton*

rums, so the network consists of overlapping communities defined by which online forum the workers use to communicate (see figure 2). But the most important fact about this network is simply that it exists. There was no infrastructure for workers to collaborate on their own, so they built this entire network on their own, off the MTurk platform, and at their own expense.

Moreover, the API that sits between the workers and the requesters rendered this network invisible. The only reason we knew it existed was because our ethnographic fieldwork allowed us to circumvent the API and observe connections between the workers we interviewed. Placing a task on the platform that asked workers to report whom they talk to allowed us to, in essence, break through the API and gather the data from the workers, which allowed us to build our map. While we mapped this network only for MTurk, our ethnographic data suggests that just such a network exists among workers of other on-demand labor platforms as well.

Today's APIs are set up to make networking seem inconsequential, yet, as in any industry, it's priceless. Those workers who aren't plugged into an active network won't see the most lucrative work and will quickly lose their footing in these markets. In fact, of the first 200 workers to do our "Facebook-Lite" HIT, almost 30 percent were connected, yet, of the last 200 workers, just over 10 percent were connected (see figure 3). Since our HIT paid well, being a part of the network conferred an economic advantage to connected workers, as they could hear about the task before the isolated workers. Platforms cut short the odds of the less experienced or more socially isolated workers simply because they don't consider why people might want to talk with colleagues to "gut-check" their options of potential employers and projects before taking a job. Top-performing workers make calculated moves based on what their peers are doing to offset this costly design flaw that dismisses the value of kind strangers collaborating to do digital labor.

While one can imagine a future technology-based fix that addresses some of the needs of workers — for instance, Amazon Mechanical Turk has since ended its practice of sending paper checks through India's postal service — the level of confidence instilled by a friend vouching for an employer, task, or platform will not be easy to replicate.

## Getting the Work Done

Poonam is in her early twenties. She and her husband, Sanjay, hope to start a family soon. After signing up for Microsoft's on-demand platform, UHRS, Poonam found she made enough money working at home to quit her job at a business processing organization and forgo the lengthy and tiring commute to the center of Chandigarh. Sanjay works full-time at a graphic design and print shop and does crowdsourcing on the side. His online earnings are an important part of the couple's income, but Sanjay sometimes finds himself struggling to manage both his on-demand work and his design job.

When we sit down with them, the couple sheepishly acknowledge that they share tasks, a risky practice. Companies expect the user whose name is on the account to be the sole person doing the work. They know that sharing work may get their account suspended, but Poonam and Sanjay take a calculated risk and do it anyway. They explain that, because they divide up the work, whoever they think has the best skill set for the task at hand can take the job. Sharing a single account helps them keep a high reputation score and doubles the number of jobs they can tackle with confidence. For instance, if one of them accepts a task but doesn't feel confident in his or her ability to complete the work, they call upon the other one. Thanks to his expertise in design, Sanjay excels at tasks that are highly visual, whereas Poonam does well with language-based tasks, like search-query evaluation.

Although one can easily imagine how common this type of coordination is among married couples like Sanjay and Poonam, we also found the practice common among workers who met on Facebook or other forums, via chat, or in person. Again and again, we heard stories about how people meet to reduce the transaction costs of different jobs, including sharing details about how to manage one's time completing tasks, how to do search queries, and how to handle tasks that require the execution of basic scripts or simple computing techniques, like copying and pasting.

For instance, Anand, 24, is a student working on MTurk and living with his parents in Chennai. Anand uses his income from MTurk to pay for personal expenses. He told us that his parents don't understand what he is doing in front of the computer so much, but he hopes to prove them wrong by eventually getting a formal position with Amazon.com. He learned about MTurk from his friend Raja, who is somewhat of a pied piper figure among Anand's friends. "Because of Raja, all of us have learnt MTurk," he says. Anand shows us the handwritten computer shortcuts and key commands Raja gave him. He keeps the tattered note taped to the wall to the left of his desk. On it are commands for saving screenshots, common search queries, and a description of how to download and search Excel spreadsheets of U.S. states and cities — a cheat sheet of sorts so that Anand can quickly answer basic questions about the U.S. that pop up when he is in the middle of a task.

Anand and Raja are not unusual. Of the thousands of on-demand workers we surveyed, on the four platforms we studied, between 5 and 10 percent reported that they tapped their social network, whether informally or directly, to ask someone for help. Workers consistently described relying on finding someone — a mentor or group of friends online or off — who was willing and able to walk a new worker through the disorienting world of survey questions, culturally specific words like "twerking," and the dizzying range of appliances and other consumer goods relatively unknown to or uncommon among India-based on-demand workers, so they could complete the most mundane tasks posted to the platforms.[7]

Fareed is yet another example of someone who relies on collaboration to get work done. A devout Muslim in his late twenties, Fareed lives in Hyderabad and supports his family by working on MTurk. His childhood friend Zaffar introduced him to the site. As the family's eldest son, Fareed often feels pressured by the older men in his family to find a more secure job. Both his father and his uncles work as drivers for hire in the United Arab Emirates, a common position for young Muslim men from Hyderabad, especially those without a chance to pursue higher education. But Fareed wants something more for himself. He joined MTurk in 2011, in part as a means to improve his English and, therefore, his job prospects. He has tried to get worker accounts on other sites but, as of our interview, had not succeeded.

Like many on-demand workers, Fareed is deeply invested in maintaining his worker reputation score. Every worker on MTurk frets about their score, because it will determine their eligibility for future work on the site.

Fareed, like a lot of on-demand workers, struggled to keep his score up at the beginning because he was getting his footing on the site. He wasn't sure how to get clarification on tasks, and even if he was able to ask, the time it took him to get answers would have taken away from his bottom line. "Rejections used to happen more [when I first signed up]," he says, "as I didn't know of [what] requesters [expected] and [understand] the given instructions for the tasks." To help raise his score, he turned to Zaffar for help finding requesters who were known to respond to workers' queries for clarification about a task's instructions.

Fareed also asked Zaffar for guidance. For instance, Fareed had lots of questions about how to do image tagging, a common task that involves choosing evocative words to describe an image. He also needed help doing location verification tasks, which meant finding physical addresses of places he had never lived, organized by streets, names, and postal codes completely unfamiliar to him. Zaffar urged Fareed to join Facebook forums — some open and some closed — created by other Indian on-demand workers. Fareed followed Zaffar's advice and forged friendships with roughly 150 people on these forums. He says, "Members share experiences . . . and when there is good work posted, [close friends] will give a 'miss call' — hanging up before the call connects saves the cost of phone conversations. We hurry to open the system and look for the work. Anyone who sees work posted calls. Whosoever is alert and sees good work tells everyone. Everyone helps everyone else."

Over and over again, workers told us that they were on the brink of giving up on ghost work platforms before they found the forums. They say it would be impossible for labor platforms to operate if workers weren't collaborating behind the scenes and that they wouldn't survive the grueling aspects of the work without the connections they forged.[8]

Fareed says that when he and his friends see a job that seems too good to be true or see a requester with a complicated job, "we turn to Facebook to ask, 'Did any of you work for this requester?' If a friend had a good experience [or more details about the task], then we'll pick up the work. Because requesters' replies [to questions about the details of a job] don't come immediately . . . asking friends is easier."

Recall that to build our network map (figure 2), workers reported who they communicated with and *how* they communicated with them. About 59 percent of all workers and 83 percent of connected workers reported that they used at least one online forum. In fact, 90 percent of connected workers communicated through a forum and 86 percent communicated exclusively through forums. Each of the dense clusters of connections in our map corresponds to one online forum, and these communities have sparse connections between them, consisting of people who participate in two or more forums (see figure 2).[9]

While most communication occurs over forums, workers also reported communicating one-on-one via in-person discussions, phone calls, emails, text messages, instant messages, video chatting, and other channels (see figure 2). Overall, 14 percent of those connected to just one other person communicate, at least partially, through one-on-one channels, and 10 percent communicate exclusively through one-on-one channels. The three most popular of these communication channels were instant messaging (27 percent), in-person discussion (18 percent), and email (16 percent).[10] We'll return to *what* workers were talking about over these different channels at the end of the next section.

## Re-creating the Social Side of Work

On-demand platforms operate as though individuals can do their work effectively without any support from others. Not a "thank you" or "good job!" or "have you thought of doing that differently next time?" But these unexamined, pernicious assumptions unravel when you talk to on-demand workers. Workers re-create social work environments to encourage one another's progress and development. They mingle in online forums that function much like the brick-and-mortar break room. And they empathize with, commiserate with, and confide in their peers.

Akbar, 19, lives in Hyderabad. He is a cricket enthusiast. He has worked on MTurk for two years and often swaps information about requesters when he sees his local friends at the neighborhood mosque. "We discuss MTurk for five to ten minutes, sharing who worked on what and what jobs were good and what jobs weren't... Later, when we go back to work... we keep chatting on Skype or Facebook. We talk and even video chat." Akbar and his fellow workers also help to keep one another motivated and awake. On-demand labor is an industry driven by U.S. and European time zones, which means most Indian workers work nights. To fill the long, dark early-morning hours, Akbar turns to his friends. "If you have to work throughout the night," he says, "you plug in earphones, put the phone to charge, and talk all night." Akbar spends most of his time talking with his friend Mohsin, because "we both use Aircel [mobile phone service] and it's free to call between Aircel numbers, so we talk a lot."

While U.S. workers rely heavily on forums to communicate with one another (91 percent of connected pairs of U.S. workers communicate with each other on forums, and 88 percent exclusively so), international workers tend to use one-on-one channels much more often (77 percent of connected pairs of international workers communicate through one-on-one channels, and 57 percent exclusively so).[11] And these patterns of use and their links to geography fall in line when we consider other, relevant pieces of context. The first is that web-based forums are nothing new to U.S.-based workers. Even if subreddits are a relatively new (somewhat toxic) scene, Usenet groups for asynchronous discussion threads are almost as old as the internet itself. But until the recent introduction of Facebook and WhatsApp on mobile devices, India had little practice in turning to the web for conversation. Arguably, the linguistic diversity—in some cases rivalries—within India may work against defaulting to online groups to organize collaboration. And finally, as much as the discussion forums could bring workers together, they also become spaces to shore up boundaries. Criticisms of "illiterate Indian workers stealing jobs" cut across all the MTurk forums dominated by presumably U.S. workers. In sum, those workers in figure 2 who have links but are not part of the largest connected component are predominantly international workers communicating via one-on-one channels, and those workers who are part of the largest connected component are mostly U.S. workers communicating via online forums.

Workers talk about the nitty-gritty of doing on-demand work, such as which tasks to do and which employers to trust, but they also provide one another with social support. Returning to the network map in figure 2, when a worker reported that they communicate with another worker, we didn't just ask *how* they communicate with that worker; we also asked *what* they communicate about, with choices including (a) HITs, (b) requesters, (c) scripts/tools, (d) life, and (e) other. Workers are far more likely to talk about tasks on online forums than they are using one-on-one means of communications. Conversely, workers are much more likely to talk about life and provide social support over one-on-one means of communication (see figure 4). The more intimate a connection between workers, the more likely they were to talk about personal matters, leaving the broadcast jumble of a discussion forum behind.[12]

## The Effects of Coordination

The API, which governs interactions between workers, has no built-in way for workers to communicate. And even though Akbar, Fareed, and

Zaffar live within a few kilometers of one another and would benefit from sharing space, the young friends work separately on their own broadband connections and cellular data plans, because they worry that if they were to share the same internet connection, they might have their accounts suspended. They didn't know if this was true or simply rumor, but they weren't willing to take the risk—an example of how a lack of clarity leads people to devise complicated work-arounds.

The lack of clarity about how and why an account might be suspended creates confusion and takes a toll on well-intentioned workers who are acting in good faith. To these workers, the account suspensions can seem arbitrary. What's more, the lack of transparency leads workers to circulate myths and theories about what prompts the platforms to ban a worker. This was a shared condition of on-demand work that could, at times, bring workers together.

One effect of the existence of the network of workers is that it allows communication and therefore myths and rumors to proliferate. Another effect of the network is that it allows the workers to, either intentionally or unintentionally, coordinate. To see this, we asked workers to do an extremely simple task—placing a pin on a Bing map indicating where they are at the moment. After they placed their pin, we asked them one question: "How did you hear about this HIT?" The answer options were (a) by searching the MTurk website for tasks, (b) from an online forum, (c) from a personal referral, (d) by searching for this requester, and (e) other.[13]

We let this task run for five weeks, and during the overwhelming majority of eight-hour periods, fewer than 50 workers did our task. But there were a few eight-hour periods in which more than 100 workers did our task, and sometimes more than 200. If we restrict our attention to the traffic spikes—the eight-hour periods with more than 100 workers doing our HIT—we see that 55 percent of the traffic came from an online forum. In fact, for each spike we saw, we found a post in one of the forums that occurred at roughly the same time. As we've seen, these forums provide a mechanism for individual workers to collaborate and provide social support for one another. They have another effect at the system level, as they cause these traffic spikes that we see in figure 5. These spikes emerge from the coordinating function that these online forums serve. In fact, roughly 200 of the 400 workers who did the HIT right after we launched it came to our task from an online forum. Note that this initial flood of workers to our task is a direct consequence of the hypervigilance of workers. Next we'll see how these networks allow workers to coordinate in a much more purposeful manner.

## REIMAGINING COLLECTIVE ACTION

Kind strangers can build connections and social space. But these communities can also create barriers to entry and keep people out. As noted in chapter 2, organized labor movements and unions historically depended on two factors to rally workers. They made use of the coherence of a professional identity — machinists, steelworkers, teachers — to lay the groundwork for a common cause. Union organizers also relied on the solidarity of face-to-face interactions at the work site and, when striking, the power of collective action as a force for change. Social solidarity, as seen from the earliest days of union organizing, could also include xenophobia and insularities. Ironically, the homogeneity of a full-time workforce and the tacit, unexamined discrimination that can make workplaces hostile to difference led some people to try on-demand work. Yet these real and imagined divisions are not hard-and-fast. Sometimes workers drafted off their professional social networks around the globe to advocate for one another as a class of workers struggling for better working conditions. One example that gained worldwide attention was the 2014 Christmas letter campaign directed at Amazon CEO Jeff Bezos.

The campaign started in the fall of 2014. Several MTurk discussion forum leaders, working with graduate students and faculty at Stanford and the University of California at San Diego, were already building a virtual union hall of sorts for MTurk's international worker base. The goal was to make it easier for workers to anonymously share stories about their experience and discuss possible actions that they could take. The site, nicknamed "Dynamo," was part of a larger research project meant to explore how to "build systems that support collective action online."[14] One of its most successful actions was the drafting of a list of guidelines for academic researchers. University students and professors have long been a source of frustration for MTurk workers, because so many of them try to recruit those workers for experiments, as though they were the equivalent of undergraduate students happy to participate in research for free pizza.

One feature of Dynamo was that workers could "upvote" suggestions, meaning a radical idea could quickly be pushed to the top. And that's exactly what happened. In short order, the suggestion of using the site to write to Jeff Bezos started trending. The emails could be fan mail or missives detailing how MTurk workers wanted to improve their work environment. The idea was fueled by Jeff Bezos's repeated and public claims that he wasn't happy if his customers weren't satisfied with their Amazon experience. He'd even offered up his email address in a *Business Insider* interview. So why not use customer service logic to advocate for the often less visible "customer" on MTurk—the worker picking up tasks for hire?

Workers logging on to the site were encouraged to submit a letter to Bezos. The pop-up note read: "The intent is to get Bezos to see that MTurk workers are not only actual human beings, but people who deserve respect, fair treatment and open communication."[15] The campaign focused on three goals: create a channel of communication among the legions of invisible MTurk workers around the world and the highly visible CEO of the company managing their work site; prompt Bezos to respond to workers' individual emails and do it in a way that made those efforts to talk with Bezos public; and advocate for specific resources for workers, though the group never put forward specific requests.

An interesting tacit goal was to correct media representations of

MTurk workers. "We don't all live in developing countries," the campaign noted. "We're not all unskilled; We're not all making $1.45/hour; We're not all Turking for beer money" — the last point referring to a commonly circulated myth that people doing tasks on MTurk aren't really workers at all. The campaign generated dozens of letters from around the world and received international press coverage. Bezos himself never replied.

In the absence of a clear professional stake and ground in which to plant it, on-demand workers tended to reject the idea that they could effectively collaborate en masse or unionize to fight for collective interests. As one worker put it, "How would we hold each other accountable for striking?" (Which is to say, where would organizers assemble to stop work?) Others felt squeamish about calling for items like a minimum wage, knowing that on-demand markets call for a global workforce. The realities of labor arbitrage — cheap labor somewhere else — meant that there would likely always be some company offering work to someone who, understandably, was willing to take a lower wage. Yet as the example above illustrates, these real and imagined divisions are not hard-and-fast. There is good reason to believe in the political possibility of organizing on-demand workers around the globe. The scaffolding and earnest collaboration in this case shows the capacity of workers to amplify their collaboration and direct it toward advocacy for themselves and one another.

## Watercooler 2.0

On its surface, on-demand labor looks atomized, ephemeral, and contingent. But upon closer inspection, it's clear that digital labor depends on workers collaborating to shore up one another's support and participation. We found that on-demand workers collaborate to reduce overhead costs, to get work done, and to re-create much-needed social aspects of work. We even found them using their professional social networks to collectively organize. Collaboration is more than an immediate, pragmatic attempt to compensate for a broken technical system. Rather, it reflects the value that people assign to social connections in work environments and their willingness to tenaciously build them into their work lives, even when it works against their earning potential. The kindness and collaborations of workers helping one another are, arguably, the digital economy's most valuable ingredient. Yet most ghost work platforms seem determined to delete it.

Purposefully or not, creators of on-demand labor platforms take aim at workers' social worlds by presuming that workers connecting is a waste of time that slows things down and is therefore worthless. The mantra of "less talk, more work" through the magic of automation seems synonymous with "scaling up" and, ultimately, market dominance in delivering goods and services on demand. Platform designers assume that better-matching algorithms, atomization of work, and the complete elimination of hands-on management are key to reducing the search costs that come with labor markets, for both buyers and sellers.

We argue that companies cannot eliminate workers' desire to invest in their jobs as something beyond a payment transaction. No system can erase workers' needs for connection, validation, recognition, and feedback without damaging both the worker and what they produce. The widespread and varied reliance on collaboration we discovered — from friends who call one another when a good job comes online to a husband-and-wife team who split tasks according to their strengths to a worker who makes a cheat sheet of quick key commands for a friend who tapes it on the wall next to his desk — all speak to the social fabric that binds work relationships.

Rather than resist worker collaboration, platform designers might want to consider taking "scaffolding" tips from the workers we met. Where collaboration works against the desired outcomes, designers could focus on explicit directions banning collaboration but build in more instructional transparency where the task quality depends on it. And when collaboration might be a boon, they could help workers do it with both peers and those hiring them to accomplish a project. In the short run, we could develop systems of scaffolding management that turn affirmation and encouragement into doable, paid tasks. We currently associate managing or curating a workforce, through recognition and coordination of collaborative teamwork, with full-time employment. To make such recognition and validation possible, not to mention training and other forms of formalized mentorship, we will need to redefine "independent/freelance" workers to better incorporate and value managing and curating digital workforces.

Collaboration put a man on the moon and built the largest, most comprehensive, most accurate encyclopedia in history: Wikipedia. Platform designers are leaving this enormous potential untapped by not building any infrastructure for workers to collaborate. Platforms like MTurk even discourage collaboration. There are precedents for collaborating online, most notably massively multiplayer online role-playing games (MMORPGs) like *World of Warcraft*. Adding in collaboration could be as simple as allowing it, providing a chat room for workers to talk to one another, and workspaces for workers and employers to confer. More complex solutions might involve shared documents and spreadsheets or shared online workspaces in general, like a virtual whiteboard.

While workers can re-create the systems of collaboration needed to find and complete tasks and build the social bonds that make work manageable, the next iteration of on-demand labor would be best served by sharing the responsibility of explicitly valuing and building out the means for workers to support one another. Work, after all, is as much a social system as a technical one. Work requires as much attention to the cultural needs and values we attach to labor as the tools to do jobs efficiently. We all stand to benefit from learning how to align the range of motivations animating on-demand workers with the equipment to help them help one another as they make their way through this demanding work.

# 6

# The Double Bottom Line

## Software, at Your Service!

This chapter explores how founders of some on-demand companies design their businesses to prioritize workers and, in doing so, create what's called a double bottom line.

Platform companies, as the taskmasters of ghost work, have two viable business options. They can sell software as an intelligent service that connects consumers who want to buy and sell what they need from one another, whether it's takeout ordered via a mobile app or web design requested from a coder on Upwork. Or companies can market people's creative insights and labor as the valuable engine behind their services. But it's businesses designed around workers' contributions, prioritizing their schedules, project interests, and collaborations, that convert ghost work into a sustainable enterprise. Though these businesses strive to "do well by doing good," they have learned, through close calls that nearly shuttered their startups, that paying attention to the workers in the loop improves their final products and services and, ultimately, their bottom line.

A good example of an on-demand, for-profit company that prioritizes workers' needs is CloudFactory. Founded in 2011, CloudFactory is headquartered in Kathmandu, Nepal, and has 125 full-time employees and a network of more than 3,000 local on-demand workers. CloudFactory offers a host of staffing support services for other tech companies. For instance, by using CloudFactory's workers to do tasks like scanning receipts and managing databases, tech companies can shift their labor costs to build revenue through sales and engineering new products and services. CloudFactory's founder and CEO, Mark Sears, could define his business as strictly a software-matching service the way other companies, like Uber, do. Legally, it is fair to say that his company matches workers to businesses in need of help. He could have washed his hands of any responsibility to those workers. But Sears made a decision to see CloudFactory's workers as key features of his business who are as valuable as, if not more valuable than, his software.

So in 2015, when a 7.8-magnitude earthquake struck Gorkha, an eastern district of the Himalayan nation of Nepal, 50 miles outside Kathmandu, Sears knew what he needed to do next. Sears, his full-time staff, and his Nepalese on-demand workers turned CloudFactory's headquarters into a crisis relief center for his workers, their families, and surrounding Kathmandu neighborhoods. CloudFactory also launched a GoFundMe campaign and raised nearly $110,000 for local relief and outreach efforts.[1] He and his team documented their earthquake relief work on the company website, to keep attention on the quake victims long after press coverage of the tragedy had ended. Like most companies, but especially on-demand companies, CloudFactory had no legal obligation to help its workers. But for Sears, CloudFactory's community of on-demand workers were the true value of his company. He saw them as a "commons" of talent that he drew from to run his business, rather than a fungible raw material to use and discard. By tending to and prioritizing their needs, ranging from safe housing to medical care, he was investing in what is called a "double bottom line" — making a profit while pushing for social change.

So what does the more common approach to on-demand work look like? The single bottom line currently dominates as a business model where on-demand companies are in dogged pursuit of opportunities to maximize profit, a totally acceptable and understandable goal in a competitive marketplace. Consider the story of Caviar, a company that sells software to connect customers, restaurant owners, and food deliverers. Consumers love the convenience of food delivery, but many local restaurants can't afford to dedicate a staff person or two just to deliver food from the restaurant to its customers' doors. Food delivery is a surprisingly complex task, requiring human-level problem-solving skills, ranging from how to keep the food warm to what to do if the customer's house number isn't clear from the street.

Caviar didn't have to look hard to find people willing to deliver food when it expanded services to Philadelphia. Sparrow Cycling, a bike messenger cooperative, had plenty of members who needed extra money and could run food deliveries in between other jobs. The bike-bound messengers who deliver food for local restaurants are part of the range of ghost work explored in this book. Caviar's couriers operate as independent contractors. The company requires deliverers to sign liability waivers for work injuries, including death, before they can start delivering.

Pablo Avendano was delivering food on his bike for Caviar on the evening of May 12, 2018, when he was hit and killed by an SUV in Philadelphia's Spring Garden neighborhood. A few days later, a friend of Avendano's, who had been a Sparrow co-worker, locked a "ghost bike" — a bike painted white to honor the dead — to a tree near the intersection where Avendano was killed. Nearby was a banner hung from an abandoned train trestle that declared, THE GIG ECONOMY KILLED PABLO. REST IN POWER.[2]

Avendano's friends organized and called for Caviar couriers to unionize, demanding a starting wage of $20 per hour and benefits that include hazard pay. They also asked Caviar to reclassify its couriers as W-2 employees. None of those demands are likely to be taken up by Caviar, as the law is clearly on the company's side and has been ever since William Randolph Hearst won his fight to classify newspaper boys as independent contractors secondary to Hearst's core business.[3]

As tragic as the death of Pablo Avendano is, there is no employment law that clarifies companies' responsibilities to people doing ghost work. Like workers on CrowdFlower, Amara, UHRS, LeadGenius, MTurk, and Upwork, Avendano was connected, managed, scheduled, and paid, in part, through a mix of APIs, AI, and web-based or mobile apps. As a worker, he was neither fully independent nor like a typical employee. Even more confusing to the case is that, for on-demand platforms, workers are often presumed to be simply another category of customer using the company's software.

Treating the worker as just another software customer is perfectly legal and understandable as a business strategy. On-demand platform services like Caviar are doing their job, many would argue, by making a profit by any legal means possible. By contrast, CloudFactory sees more profit in treating its workers more like business partners. Mark Sears, CloudFactory's founder and CEO, chooses to take a different tack with on-demand work. But CloudFactory's approach doesn't fall under any clear labor laws, either. A sole proprietor's intentions, whether pursuing a single or double bottom line, do not fully address the needs of those doing ghost work precisely because individual companies cannot set the rules for the entire growing world of ghost work fanning out across all industries. Typical bottom-line business strategies can't protect the interests of ghost work like the safety net that comes with formal employment. Because Caviar operates as a software company solving the macro-task of food delivery, it is not—or has not yet been—held responsible for Pablo Avendano's death.

The future of on-demand work, and the accountability and care for the conditions of those doing macro- or micro-task ghost work, is at a crossroads. One option is to hold companies that rely on software and a human labor pool to deliver a service to consumers legally accountable to that labor pool as their employers. Another option is for workers to continue to fend for themselves through the kindness of strangers and well-intentioned companies hiring them on contract. There might be other viable options that make ghost work a more sustainable form of employment. Finding our way to these other options that distribute the benefits of this new economy to workers, on-demand services, and consumers more equitably could come from considering how far intentional design focused on both profit and worker experience takes us, compared with the limits of narrowly focusing on a single bottom line.

## SINGLE BOTTOM LINES THAT CONVERT WORKERS INTO CUSTOMERS

On-demand services that sell themselves as software companies typically maximize their bottom lines by cashing in on both sides of the ghost work market. Customers paying to use an on-demand service's platform to get food, hail a ride, or clean training data are one revenue stream. The service can also make money on this transaction by marketing and selling app users' information to advertisers.

Workers generate revenue, too, as they are considered another customer paying the platform for use of its software. Platforms profit from workers' software use in three ways. First, like any broker matching a supply of workers to a demand for labor, platforms make money charging a fee each time a worker completes a requester's task on that platform. On-demand services also profit from thc valuable information that workers generate as they use the platform. The on-demand service can digest these workers' activities and turn them into training data to improve or automate parts of what their software delivers. Lastly, on-demand services can also market and sell workers' information to advertisers. On-demand services are the perfect business model for maximizing the digital economy's single bottom line.

Caviar is not the first and definitely not the only macro-task ghost work service that argues it has no responsibility to workers picking up jobs on its platform. Companies focused on a single bottom line, hoping to profit from the premise that they are solely software services, are well positioned to argue that they are neither employers nor sites of employment. As discussed in the first chapter, CrowdFlower, an early rival to MTurk, successfully defended, albeit through an out-of-court settlement, the same stance. It made its case that it was a matching service rather than the employer of record for thousands of people doing ghost work, and no court case or law has undercut its footing to date.

Much of the legal murkiness of the CrowdFlower case, like the culpability of Caviar for the death of Avendano, given that the courier delivered food for Caviar's on-demand service, is shored up by a string of cases dating back to the late 1990s. These settlements, rather than any intentionally designed law, created the de facto benchmarks used today to defend treating tech's contingent working class as something other than employees. Workers are compelled to contract out their labor at their own risk. And more and more businesses are depending on crowdsourced labor pools as they move their operations to API-managed contract staffing tucked behind the curtain of AI. *Caveat venditor.*

Today, Uber is the most visible example of a company that sees itself strictly in the business of offering people software to help them find rides from people available to pick them up. The popular mobile app offers not only "peer-to-peer ridesharing" but a buffet of what it calls "transportation networks," from carpooling and food delivery to private jets. Hardly a month goes by without a new lawsuit or settlement resetting or spinning Uber's business model.

According to Uber, rider-customers drive the transaction. Rider-customers open Uber's app and use its blend of software and APIs to find someone to hire for a ride. Uber driver-partners are, for Uber, another customer. They use the platform's software to make money by offering their time and personal vehicle to respond to rider-customers' requests, picking them up and dropping them off at a destination plugged into the app. The challenge of the single-bottom-line framing of the worker as another software user is that customers hailing a platform's software to call ghost work into action rarely, if ever, see themselves as responsible for the work conditions of the human delivering their part of the software-as-a-service.

For example, the Pew Research Center found that most customers hold a mixed, often contradictory set of beliefs about what Uber does and does not owe driver-partners.[4] Most respondents felt strongly that ride services like Uber should not have to follow the same rules and regulations as taxis. The majority of respondents also viewed ride-hailing services' drivers as independent contractors rather than employees. Yet they also expected Uber, as much if not more than the driver, to manage their customer experience. This contradicts all of the current rules we have about how much a business can manage or direct those working for them on contract and what help or support an independent contractor can receive from a business that is not one's official employer.

On April 30, 2018, the California Supreme Court ruled that some Uber driver-partners should have been classified as employees rather than independent contractors. Uber was found guilty of violating existing labor laws where it withheld employee benefits to individuals who could show that Uber controlled their wages, hours, or working conditions, ability to secure work, or a "common law employment relationship" in which the ridesharing company trained, directed, or controlled how a driver-partner carried out their tasks.[5] The litany of legal cases brought against Uber begs two larger questions: Who is the "employer of record" in a ghost work economy driven by a single-bottom-line scenario, and when are consumers of ghost work acting as employers, too?

Legal fights over how Uber classifies and treats its driver-partners as not just another customer, but a worker on its platform, make visible how challenging it is to define on-demand workers' rights more broadly. Who's accountable to workers who, technically, are employed by consumers to do something for them but do so through an on-demand company's APIs and software, operated over the internet?[6] Without question, talking with a driver-partner about their work experience makes it easier for the average Uber customer to see the strain that a driver-partner might feel under current employment classification systems. The greater challenge is that Uber is the visible tip of an iceberg. The end customer benefiting from most ghost work will never physically meet—often even know of—the people delivering a service to them. Because, of course, hiding the people in the loop from the end customer is part of the value proposition of software-as-a-service.

But treating workers like ghostly figures is not a given or necessary element of on-demand services. Some companies see themselves as not only software providers but also purveyors of human expertise and creativity. They consider it their business to be responsible for workers' labor conditions because they believe it's better for their business in the long run.

## THE DOUBLE BOTTOM LINE

Where some companies see only consumers, others see a workforce. They are also motivated by more than a single bottom line. These platforms' designers assume that humans in the loop are not going away. They are among a growing number of what are called "social entrepreneurships" that hold themselves accountable for making a profit for their investors, like any other for-profit enterprise. They also explicitly put themselves on the line to meet a measurable social welfare goal, from being carbon-neutral to expanding job opportunities. Companies from Patagonia to those profiled in this book, like CloudFactory and LeadGenius, register as benefit corporations, or B Corps, to make their commitments a matter of public record.

B Corps are for-profit entities that register their intent to make a positive impact on its workers and society at large, through policies like fair labor practices, community giving campaigns, and clean environmental practices. The utility of the B Corp label is up for debate. Some argue that a vague mission — the corporate slogan equivalent of "Don't be evil" — can water down the label's meaning.[7] But the point is to help make companies accountable to society as well as to shareholders.

In the on-demand marketplace, B Corps like CloudFactory push back against the prevailing notion that people doing ghost work are expendable. They prioritize workers' schedules, interests, and collaborations. Ultimately, these platforms show how worker-focused design can improve the quality of work produced and workers' experiences of their jobs.

## Double Bottom Line by Design

The co-founders of LeadGenius felt a personal connection to their workers. For Anand Kulkarni the connection came when he traveled to India to visit relatives. At the time, he was a PhD student studying industrial engineering operations at the University of California at Berkeley. While in Mumbai in 2010, Kulkarni took a tour of the slums in Dharavi. His guide, who lived in the slum, asked Kulkarni to help him build a website for his tour company. Kulkarni was willing but didn't know where they could access a computer and software. The tour guide did. He navigated through makeshift tea stalls and fruit stands held together by corrugated roofing and bright-blue wooden posts until Kulkarni saw, wedged among the shops, a cement room with a blue tarp for a door. Inside the cement-block structure was a cybercafé.

Kulkarni remembers that, even in a neighborhood with no running water and a patchwork of borrowed electricity, the internet access was pretty good. "Technology was there in the slum. What people didn't have was access to opportunities in their local economy." Upon returning to Berkeley, his driving question became: Could technology enable people in the slums of Mumbai to make a living if it connected them to the global economy? And Kulkarni didn't want to stop there. He wanted a design that put jobs directly at people's fingertips through their mobile phones.

Also studying engineering at UC Berkeley that year was Prayag Narula. Narula met Kulkarni when both men signed up for a seminar program called Information and Communication Technologies and Development, taught by Tapan S. Parikh, a professor with a reputation for hatching startups with social missions. At the time, Narula didn't know much about crowdsourcing or using APIs to route work, but he deeply believed in the power of the internet for sharing information.

Narula also had a personal connection to India, having grown up in a two-room house in Uttam Nagar, a rough township on the west side of the city of Delhi. As a boy, Narula had loved books, but his town didn't have a library, so he went to electronics shops and internet cafés to download pirated copies of books onto a portable pen drive. Initially he read them on a computer, but later he got a smartphone and would read on the phone's tiny screen.

Kulkarni and Narula wanted to build an on-demand service platform where workers could quickly come together to tackle a new project as a team; this is called a scaffolding approach.[8] The goal is for each team to have a mixture of newbies and experienced workers. New workers bring new questions and new ways of orienting themselves to a problem, while experienced workers cement their knowledge by passing it down. Each team was overseen by a junior manager, who was available to answer questions. This strategy is the pragmatic way to help a business grow quickly, even if Silicon Valley uses the more vacuous, depopulated language of "scaling up." Kulkarni and Narula believed that scaffolding their workers as teams could accomplish more than an unsupported, atomized universe of independent workers.[9] And, like CloudFactory founder Mark Sears, they believed that workers with support produced better work and were ultimately better off. So, to create LeadGenius's first worker base in 2010, they started with the most obvious place to find people connected to one another—Narula's cousins, uncles, and aunts back in Uttam Nagar.

## SCAFFOLDING AS A BUSINESS STRATEGY

As noted earlier, LeadGenius's software integrates workers into each step of evaluating prospective sales for the company's clients. Workers don't just chase leads; they check one another's work, they cheer each other on, and existing workers recruit new people to join them. Teams working on specific client campaigns keep tabs on each other's work and collectively identify when the sales-lead data are too thin to pass along to a client, deciding as a team what to do next.

At first, Kulkarni and Narula braced themselves for high turnover in their workforce, assuming that workers would absorb the training and experience gleaned from LeadGenius and then use it as a stepping-stone to traditional employment. But they were relieved to find that their workers stuck around. Purposeful design choices they had built into LeadGenius's basic platform operations, like real-time chats and worker discussion boards, created a lively and welcoming community. The company furthered the sense of teamwork and collegiality by keeping teams connected to the same projects—what they call "campaigns"—until completion. Because campaigns last anywhere from one week to several months, workers get to know one another, especially given that they have use of the platform's live-chat software. The live chat allowed conversations to bubble up organically, like watercooler conversations might in a brick-and-mortar company's break room. And if a worker wanted to take time off, that was an option, too. Again, the worker's own interests and schedule were given full consideration, which made for a more engaged workforce.

Normally, a company like LeadGenius would have three types of workers—experimentalists, regulars, and always-on—who reflect the Pareto distribution. But LeadGenius made a conscious decision to require workers to put in at least 20 hours of work a week on the platform. As a result, LeadGenius ended up with a blend of workers who were always on and regulars. The more consistent level of experience meant teams were better able to produce high-quality work quickly. The formula seems to be working: since picking up a stable client base, LeadGenius has been able to pay out more than $5 million a year to its workers.

By using this scaffolding technique, LeadGenius reinvented a career ladder. Rather than leave people to individually absorb the costs of orienting to new projects or figure out how to move forward when they got stuck, LeadGenius invested in capturing the value of all the collaboration described in the previous chapter as the kindness of strangers. They made paid tasks out of mentoring others, supporting teams, training new hires, and the hiring process. One of the greatest differences in LeadGenius's design is that it assumes that people may want or need to move back and forth between contributing as an always-on worker and a regular one, much as Zaffar did when his mother was in an accident and he needed to step away from LeadGenius for more than a month to care for her. It sees workers' desire to control their time, what they work on, and who they work with as features rather than bugs of on-demand systems.

LeadGenius also assumes that each campaign's focus, rather than the skills of a specific subset of people, will determine the mix of team members assigned to the job. This design allows them to draw on a worker's personal interests as much as a worker's past successes. If, for example, someone is a sports fan, they may become a fantastic worker on a project looking for sales leads related to selling software for tracking fantasy sports leagues. LeadGenius workers can select projects by opting in on campaigns posted to weekly job message boards. And unlike with open-call platforms like Upwork and MTurk, workers are not forced to vie for jobs in ways that might drive down fellow workers' pay.

LeadGenius was one of the first on-demand businesses to build in-house tools to help workers collaborate. They didn't assume, like many of their competitors, that workers' communication was bad for business. Some of this comes from the realities of managing a global workforce operating in 45 different countries: communication is vital when coordinating across time zones. They addressed this, in part, by organizing teams by country, led by junior managers located in that country. Once the junior managers were connected, it was easy to connect workers as well. Open chats among workers create cohesion among team members, which is both a business advantage and an advantage to workers, who can ask peers for assistance if they get stuck on a problem.

LeadGenius relies on its most experienced members to recruit and hire people like them. The company provides software and tools, and in some cases even a tech budget, to equip workers to contribute. "It makes everybody's lives easier," says Narula. "And it helps us by increasing the length of time workers stay with us." For Narula, this is the model design for the future of work and the only sustainable way to use ghost work for more complicated workflows. He believes that learning from how workers organize themselves — how they make decisions, manage their time, and collaborate on projects—strengthens the work delivered.[10] "As we provided more and more tools, that self-organization became stronger and stronger and the results got better."[11] While LeadGenius offers a for-profit model that prioritizes workers, a nonprofit approach can afford to elevate workers' needs even further.

## WHEN VOLUNTEERS BECOME WORK TEAMS

Imagine an on-demand platform that paid you to do something that you already loved to do. That's how the nonprofit Amara started its on-demand captioning and translation services.[12] Those roots influence every aspect of the platform today, but they can be found from its earliest community projects.

Back in March 2011, people used Amara to translate and share information during the Fukushima Daiichi nuclear power plant meltdown, in Japan. That same year, Amara played a role in information sharing during the height of the Arab Spring. Then, in late 2011, TED, which had its own robust community of 7,000 volunteers from all across the globe, approached Amara with a proposal. TED's leaders felt Amara could help them meet their intent to quadruple the size of the TED community and increase the number of languages used for recorded talks archived on their site. Today, TED and Amara coordinate a large volunteer community called the Open Translation Project. Using a version of Amara's tools, overhauled with the detailed feedback from the TED translators and captioners, the TED OTP (now called TED Translate) includes more than 50,000 members who have translated over 100,000 talks into more than 116 languages.

Back in the early days of Amara, most volunteers used the platform to translate and caption videos for the TED Foundation's Open Translation Project. Many had personal reasons for volunteering their time. Some people loved a particular TED talk and wanted more people to hear it. Others felt strongly about accessibility and hoped to remove barriers preventing the deaf and hard of hearing from enjoying TED. At first, money was not a part of the platform. On occasion, an organization would ask Amara to subtitle its materials quickly or feature its projects on the Amara community homepage. But that's not how Amara worked. Typically, requesters had to wait for a group of volunteers to coalesce around a project that excited them and then caption it. It was only a matter of time before a business was hatched.

In 2013, a film distribution company came to Amara with a need to translate roughly 50 films into 18 to 22 languages each for several international film festival deadlines. The film distribution company had searched online for a service that could subtitle all of its films, but the distributor's grant would, at best, cover only the costs of subtitling a small percentage of the languages in which they wanted to screen the film. The company then found Amara and asked if it would take on the job. Amara said yes, believing it could find enough translation professionals within its large volunteer community to manage the task. Within weeks it became clear that the size of the task would require more Amara community members for the job. That was when Amara's leaders, led by tech strategist Aleli Alcala and then PCF Executive Director, Nicholas Reville, launched Amara On Demand (AOD). They turned to the site's large volunteer community of professional subtitlers, posting a note to the homepage asking if anyone would be interested in getting paid to do some caption and translation work on a tight deadline. They let people know the challenge at hand, including the limited budget that would dictate the pay rate for the project, and the impact their help would make toward ensuring these powerful documentary films made their way to the prestigious film festivals that wanted to share them. The call resulted in an outpouring of interest. Soon Amara had roughly 200 people interested in being paid for work that they had previously done for free. It was a globally dispersed, passionate group who were excited about helping a nonprofit film distributor get content out in as many languages as possible. Even though the work was paid, the Amara On Demand Team, as they came to call themselves, still operated with a feeling that this went to the heart of Amara's mission of accessibility and inclusion.

Dean Jansen, Amara's executive director, and Aleli Alcala, the organization's chief operations and sustainability officer, decided to focus on building out Amara's business opportunities and the Amara On Demand model for fair, sustainable work. They looked at companies like Uber, and on-demand platforms like Amazon Mechanical Turk that kept popping up and getting traction. The income gaps between the executives building these companies and the workers carrying out the services only seemed to be widening. Venture capital funding models didn't necessarily help close these disparities between company founders and workers hired to perform on-demand services. Alcala and Jansen saw Amara as an opportunity to offer a different future. Jansen describes Amara's transition from an all-volunteer community to an effort to provide sustainable, on-demand jobs, saying it was "kind of like a horror movie" when he realized how hard it was to fund a fair on-demand employment business model in Silicon Valley.

Amara had a small paid staff and lots of volunteers, but it didn't have a business operation that fit the world of on-demand work. Alcala and Jansen wanted to make the organization financially sustainable. And they both believed in the importance of running Amara as a nonprofit so that the organization was not beholden to angel investors or pressured to be anything but committed to Amara's mission. They wanted to build Amara into an equitable organization and business, despite the uncertainties of the future of work. Rather than chase venture capital seed money, Amara's founders chose to focus on funding their nonprofit startup the old-fashioned way. They pitched AOD to business clients who wanted to work with a mission-driven organization that could offer the higher-quality work that Amara's on-demand services produced, at affordable prices.

Amara offered paid work alongside its volunteer opportunities. Each job provides decent pay and enough hours to support the base of team members contributing to the project. Amara functions as equitable and fair work in an online setting. It's focused on developing ways for people to shape the governance of their workplace, pay rates, and job opportunities. For example, when Alcala schedules large projects, she factors in AOD members' family vacation plans, kids' birthday parties, and school obligations. Those are part of her priorities when putting together the AOD work schedule.

In 2015, Amara broke even with earned revenue for the first time. They did it by selling the value of their double-bottom-line strategy. The on-demand team that started with roughly 200 members now numbers upwards of 3,000. An average of roughly 350 people a month are paid to do captioning and translation work on Amara. Reflecting the Pareto distribution, roughly 10 to 20 percent of the people affiliated with Amara do 80 to 90 percent of the work. Amara's clients range from big technology companies to online education learning platforms, and some of the Amara On Demand (AOD) linguists are starting their own businesses and becoming partner providers working with Amara, something like franchises of AOD. There's an ecosystem that's growing around Amara On Demand. What sets Amara apart is that it continues to offer people volunteer opportunities alongside paid work. At any point, someone can offer their translation expertise to a cause that means more to them than the money.

## The Range of Ghost Work Operating with a Double Bottom Line

There are other models of the double bottom line worth considering. One is a focus on explicitly connecting workers and the customers they serve, exemplifying what is called a peer-to-peer sharing economy.[13] For example, in 2015, two Bay Area entrepreneurs launched Josephine, a company that connected people who loved to cook with people who wanted home-cooked meals.[14] A primary selling point of the service was a sense of community. Instead of the impersonal delivery that comes with Caviar, people who used Josephine picked up their orders at the homes of the cooks, connecting people in the intimacy of a family's kitchen. Most of the cooks who signed up were already making meals for large families, which made it easy to set aside an extra plate or two of food. Some were recent immigrants to the United States who wanted to share their favorite recipes and childhood food traditions. The cooks posted menus of home-cooked meals on Josephine's site, and customers signed up for the meals they wanted to buy, then picked up their order in person. Josephine took 10 percent of the sales. At its height, Josephine had roughly 75 cooks enrolled in the Bay Area. It received accolades from the press for its progressive mission to connect people, many of whom were already neighbors but didn't know one another, through a shared appreciation of a home-cooked meal. In many ways, Josephine was an ideal example of the sharing economy meeting the on-demand economy — foodies who were passionate about cuisine connected through APIs and software with people who appreciated the effort and quality that goes into a home-cooked meal.

## CREATING A SEAL OF APPROVAL

Another strategy for making sure workers' needs are met is to develop a fair-trade market for good ghost work practices. Given the choice and the financial means, some people want the option to be responsible consumers of other people's labor. With the explosive growth of online platforms built around marketing the labor of domestic workers — nannies, housecleaners, and caregivers — came concerns about how they and other service sector workers would be treated. Online companies were running into the same issues as their more traditionally run peers, meaning the domestic workers they managed were starting to struggle with low wages, unstable income, and a lack of benefits. The challenge became how to reward online companies that set a higher bar for employing workers and how to signal to consumers that a company was worth paying a little extra because that money supported labor-friendly policies. The Good Work Code was designed to answer those questions.

Created in 2015 by Palak Shah at the National Domestic Workers Alliance, the Good Work Code guides tech companies in the creation of jobs that are sustainable for on-demand workers. Here's how it worked: By signing on to the Good Work Code, companies were committing to eight core values that made them friendly sites for on-demand workers: safety, stability and flexibility, transparency, shared prosperity, fair pay, inclusion and input, support and connection, and growth and development. In exchange, the companies got to use the Good Work Code label in their marketing materials as a way to signal to consumers that the company offered a supportive and sustainable labor model for its workers.

An example of the Good Work Code in action is DoorDash, a food delivery service. DoorDash carries a commercial auto insurance policy that covers up to $1 million in bodily injury and/or property damage of its Dashers, who are considered third parties. Any accident that occurs while on an active delivery (in possession of goods to be delivered) qualifies the Dasher for coverage. Dashers set their schedules, and DoorDash maintains a list of best practices and tips on "how to dash." Though the service is careful to point out that it doesn't presume to tell workers how to do their jobs, the implications are clear: *Follow these tips if you want to keep dashing.* The guidance walks a fine line between offering advice and telling Dashers how to do their jobs, a breach of independent contract workers' employment rights. And DoorDash is one of the first on-demand platforms to offer perks to workers, akin to the benefits of more traditional jobs. They offer services like Stride through third-party providers, to help Dashers sign up for the Affordable Care Act. DoorDash partnered with another on-demand company called Everlance to offer Dashers help tracking their business expenses, so that on their taxes they can get their fair share of the deductions that come with being self-employed as independent contractors. And DoorDash offers a same-day pay feature, called Fast Pay, that lets Dashers cash out earnings, much like Amazon Payments lets workers draw their pay into a direct deposit account. Fast Pay charges $1.99 to cash out. By aligning with the Good Work Code, DoorDash hopes that employees will be happier and want to stick around, which will translate into more customers and, ultimately, a healthy business.[15]

## CO-OPS 2.0

A last model to consider is the burgeoning "platform cooperativism" movement. Cooperatively owned, democratically governed businesses hark back to the guilds that predate the Industrial Revolution. Cooperatives allow workers to operate as shareholders distributing costs and labor equally among worker-owners. Platform cooperatives use software and APIs to facilitate transactions, sales, and exchanges of goods and services. They can even work hand in hand with labor organizing.[16] Labor unions are a natural source of support for platform cooperatives, as both have an interest in workers' rights.

One of the best examples of a union joining forces with a platform cooperative is the NursesCan Cooperative, in California. In 2017, the Service Employees International Union–United Healthcare Workers West (SEIU-UHW), a 150,000-member union, threw its support behind the NursesCan Cooperative, a platform of licensed vocational nurses who provide on-demand care. In the hope of creating more stable, better-paying jobs for nurses, the union offered legal support and connections to potential employers. The idea behind NursesCan, formed by five licensed vocational nurses, was to deliver on-demand, at-home care for patients who wanted the option of a nurse visiting them at home, as opposed to going to a healthcare facility to be seen by a healthcare provider. As the population ages, the need for nurses in the United States alone (not to mention the rest of the world) is projected to grow 12 percent between 2016 and 2026, according to the Bureau of Labor Statistics.[17] The need for more on-demand services to coordinate skilled nursing care seems inevitable. But healthcare is one business that needs to support regulars and always-on workers to provide consistent coverage. Most consumers will want more than experimentalists caring for their loved ones. Key to this market, then, will be guaranteeing that always-on workers can take breaks, refresh, and update their skills, and that regulars can be part of teams that help them learn and develop their expertise, with the guidance of more experienced workers. Everyone wants workers invested in the business of healthcare as much as the on-demand services that might be routing them to someone's home for care.

All of the businesses above offer examples of what it could look like to have workers factored into the center of ghost work's concerns and how doing so benefits business owners and consumers. At the same time, none guarantee that the range of workers' interests will be met, despite the good intentions of a few individual business leaders.

## When Good Intentions and Better Design Are Not Enough

Kulkarni and Narula built much of LeadGenius's workforce outside the United States when it reorganized from MobileWorks. The company could not afford to pay U.S. workers the equivalent of the U.S. federal minimum wage. Nor could LeadGenius clearly classify its workforce as contractors and beyond the requirements of federal minimum wage laws. Why? Because to build community, you need to do things that traditional employers are allowed to do only for employees. Providing training, giving people direction on how to do their work, negotiating schedules, even offering people company T-shirts were all practices that helped LeadGenius build its community. It was part of its design. But these are also features that, by legal litmus tests, could classify LeadGenius's workers, if based in the United States, as employees rather than independent contractors. The company can afford to invest in its team researchers this deeply only because it does not have the additional costs of employee benefits and payroll taxes that would accompany traditional employment. LeadGenius can afford to treat full-time and part-time workers, regardless of where they live on the planet, as, effectively, the same. The company, however, can't afford to offer them healthcare, paid leave, or social security benefits that would transfer globally. Nor can it keep national borders that sort and rank the value of different workforces and how they are compensated from fracturing its global team. Like all companies that outsource globally, LeadGenius applies the logic of labor arbitrage to set wages. It pays workers what the market will bear to have products and services "localized" — translated into a specific language and packaged with local character.

Labor arbitrage refers to the practice of seeking out workers living in locations with lower costs of living and paying them the typical local wages for the work they do. The practice is not new. Labor arbitrage is at the heart of the outsourcing that fissures the workplace, as discussed in chapter 2. In theory, a healthy, competitive market with lots of employers competing for the same workers keeps wages fair. A rising tide lifts all boats. In practice, a few big companies control the supply chains — those networks of suppliers feeding a company's production and distribution process. As such, those large companies also control the contract job prospects. Many companies in the business of information services can take their business anywhere, leaning on the power of their supply chains and the internet. Under these conditions, wage stagnation is inevitable. This can create friction among workers when they are no more than a chat screen away from peers in a different country doing the same work for twice the pay.[18]

For example, LeadGenius's workers are organized in teams that work with one another through internal chat and text systems. They are often trying to figure out who they are working with and whether those team members make more money than they do. The realities of LeadGenius's pricing and its dependency on lowering prices through labor arbitrage mean that workers can be left feeling undervalued because they find out what other countries' workers can earn doing the same work. The logic that one deserves equal pay for equal work is even tougher to refute in places where workers see little difference in the costs of living when they compare, say, India with the Philippines — as some of our interview participants did — making discrepancies in pay hard to accept. Add to this the persistent message of community and the value of teamwork and the importance of each person's contribution, and it becomes hard for a company to quell the frustrations of workers who feel they are told they are equal but do not receive the same pay.

Ironically, in creating a minimum number of hours for workers, LeadGenius also sidelines those, like Natalie, who started out on MobileWorks but did not have a set number of hours to offer to meet the minimum requirement of 20 weekly hours at LeadGenius. She and other experimentalists were entirely cut out of the opportunity to work. For other workers, the expectation that junior managers take night shifts as needed meant that new mothers, like Lalitha, and newlyweds, like Zaffar, couldn't take advantage of the only ladder of advancement available. No single employer is mandated to cover his healthcare as he starts his family. And, despite the value of being able to return to LeadGenius after his mom's accident, Zaffar had no choice but to take a leave without pay.

Despite Amara's amazing software design, it competes in a completely unregulated market. That allows its competitors to outbid them for most projects, because they, unlike Amara, pay incredibly low wages, through tasks often posted on platforms like MTurk, to save costs. It will take Amara making enough of a name for itself as a higher-quality product to cut through the low bidding of other translation services. These competitors push wages down and face no censure for exhausting or depleting the pool of on-demand labor that they in many respects share with Amara and other businesses that need to draw from a common community well of translators to run their businesses.

Both LeadGenius's and Amara's success at fulfilling their double bottom lines tacitly depend on factors far outside of their control, namely, internet access and both basic computer literacy and a liberal arts education that equips people to think creatively and learn how to learn rather than recall facts that any computer can now store. It is true that far more people around the globe have access to the internet than ever before. At the same time, after 20 years of trying to close the divide between digital haves and have-nots, the mere presence of technology is not enough to level the playing field. The persistence of the digital divide and the digital inequalities that follow from it will become an even more pressing problem if the vast majority of participation in a global information services economy depends on three things: a stable internet connection, continuous access to educational resources, and the means to cover basic healthcare, sick leave, and family leave no longer provided by a single employer.

The devil is in the details, and, in the case of Josephine, the details looked like regulations set by local and state health departments to manage large-scale industrialized food prep. California, like most states, requires that food sold to the public be prepared in a commercial kitchen. Inspectors from the California Department of Public Health sent cease-and-desist letters to Josephine's home cooks. The company tried to reconfigure its format to accommodate the laws. The founders rented commercial kitchens for its cooks and lobbied for new regulations that could accommodate consumer demand for nonindustrial food. But in the end, the new format didn't work, and Josephine suspended operations in early 2018.

The downside of the Good Work Code is that its seal of approval relies on consumers noticing that there are choices and caring enough to take the more expensive or harder course of action to benefit the double bottom line and social responsibility. Companies like DoorDash that signed on to the Good Work Code still need to cover their bottom lines. For example, DoorDash may offer liability coverage, but it also requires its independent contractors to maintain their own insurance, too, in whatever amount might be required by local laws. If Dashers don't carry their own insurance, DoorDash's coverage may not apply. Add to this the challenge of labor organizers figuring out a business model for organizing workers who do not share the same job site and may see one another as competitors instead of co-workers.

Similar tensions involving lack of shared work sites is a potential stumbling block for platform cooperatives, too. Traditional co-ops depend on everyone giving equally to the co-op. They also are locally focused, depending on a local market to buy and sell the products to create a network of peers supporting one another—providing a cooperative supply chain. But co-ops haven't figured out how to maintain a cohesive membership across national borders, particularly the nuts and bolts of how to have international members share equally in the profits. They also haven't figured out how to have equal accountability and democratic collective decision-making in a Pareto distribution where people come and go.

It turns out that people carry different mental maps to their ghost work.

Nothing illustrates this more clearly than a realization that we had after completing our research and looking back over interview transcripts. We had asked workers, "How do you describe what you do to other people?" One U.S. worker told us that he "worked for a startup in Silicon Valley." Another worker, also from the U.S., told us that she was self-employed and using ghost work to build her business skills. And a third worker, based in India, said that he was an entrepreneur bringing jobs to his village. All three people worked on the same platform doing the same types of ghost work tasks. Neither cooperatives nor labor advocates have figured out a business model that effectively bridges across the mental maps produced by ghost work.

Despite the best intentions, businesses' good efforts fall short. On-demand services lean on the natural order of a Pareto distribution, functioning more like an online community and a self-organizing commons than a factory. Developing a sustainable model for on-demand work will require more than attention to the technical design of APIs, platforms, and software. It will also demand rethinking how we shore up the economic and social development of people's capacity to participate in a global economy that's not going to be fully automated anytime soon.

## Tragedy of the Commons

A cynic might read the efforts of CloudFactory, LeadGenius, Amara, and other companies espousing double bottom lines as using the trappings of their B Corp or nonprofit corporate status to look like the heroes in an economy with easy-to-find villains. Despite the shortcomings of companies like these, they are willing to call out their responsibilities as sources of employment for tens of thousands of people around the world.

LeadGenius and Amara are two platforms deeply invested in fostering worker interaction and task collaboration. Their work practices feed a shared pool of people interested in working for any number of on-demand platforms. For example, LeadGenius added a minimum wage, set hours, what they call "scaffolding" mentorship, and advancement back into the mix. Amara allows workers to choose between volunteering and doing paid work, creating an exciting blend of people making a living and able to share their "language pride" to contribute to something bigger. Conversely, employers who are optimizing only profit focus on making workers another kind of customer or revenue stream.

LeadGenius and Amara create ways for workers to control their destinies, offering models for how to support teams that are collaborative, cooperative units. Amara, in particular, points to a possible future that puts the worker in the driver's seat, able to set their schedule, negotiate wages and profit-sharing opportunities, and make decisions about when and how to contribute time and effort to projects that they value beyond a price tag. Helping workers connect, fostering rather than ignoring or stifling their collaboration, is not just the right thing to do for workers who contribute the lion's share of labor to these systems. Providing platform-based discussion forums and offline meet-ups and rewarding more experienced workers for showing newbies the ropes will likely generate the highest-quality work and products for customers.

All the on-demand services' principal engineers and business leaders shared their doubts and crises of conscience about trying to make on-demand work both financially profitable and productive for themselves and their workers. They are doing everything they can to value and tend to the commons of ghost work labor that they draw on to run their companies. We share their concerns. The final chapter will outline the technical, political, and societal hurdles we must face to transform ghost work and improve the lives and futures of the people powering the on-demand economy.

# CONCLUSION

## The Task at Hand

### Ghost Work Under Construction

Now that the full range of ghost work — from image tagging and video captioning to debugging and food delivery — is easier to see, picture this. Imagine someone at your gym pouring himself into his running routine. His recent soccer injury flares in his left knee. He steps off the treadmill next to you. He seems to recognize this feeling. Maybe you do, too. You see the flush of swelling around the kneecap. Growing pressure overtaking his left leg as he shifts his weight to the right. You see him tap a band on his wrist three times and, in a flash, you hear his health coach say, “Hey! How are you doing? I’ve just pulled up an activity scan for you from the past hour. It looks like you’re having some physical distress. Is it your knee again?”

You hear him respond, sheepishly, “I know that you suggested this morning that I might want to shift up cardio routines this week. You were right.” Within seconds, the health coach responds, “I’ll find a physical therapist who’s worked on knee injuries like yours before.” Less than another second passes. “Okay, got someone. They’re available and can join you at the gym in 15 minutes. I’ll check in with them now and update them on your situation. Give me a minute.” “Thanks,” the man says as he leans against a stationary bike. “And we need to get that left leg elevated,” the health coach adds. “I can see that you’re still putting weight on the knee." This is the business future imagined by Twine Health, a startup acquired by the activity tracker Fitbit.

Twine Health is the almost visible end of a supply chain that delivers patient care and health coaching through text messaging and voice calls. Twine's main customer is the patient who reinjures his knee at the gym. Twine's bigger market is the corporation trying to reduce health costs through a workplace wellness program. By looping into Twine's APIs, a certified health coach can care for scores more patients than one working in a traditional office setting. A third party, a company called Wellness Coaches, feeds Twine's demand for clinicians. Today, no rules manage how companies like Fitbit and Twine operate their supply chains of on-demand coaching as they deliver patient support.

Similarly, a combination of cameras, sensors, internet connectivity, and real-time support from a health provider make on-demand surgeries easy to imagine. A team of on-site nurses and remote, on-demand surgeons will be able to perform everything from appendectomies to cataract surgeries through robotics and monitoring. Teams will coordinate closing a patient's incisions and double-checking each other's suture work. Specialists will be brought to the table to consult when a sensor picks up that a patient's blood pressure has dropped or if excessive bleeding suggests a complication.

On-demand workers could also factor into the industrial world of construction and heavy machinery maintenance, from building wind turbines to repairing oil rigs. Companies like Daqri have already developed augmented-reality-infused gear that bridges on-site and remote on-demand workers. The company's Smart Glasses, for example, include a computer chip, cameras, and sensors that can project a 3-D map or repair manuals and can scan for signs of dangerous heat and pressure buildup in piping. It can help an on-site worker patch a complicated hydraulic system with the expertise of an on-demand pipe fitter pitching in. As autonomous vehicles advance, Daqri's Smart Glasses could be the headpiece of an industrial robot augmented by an on-demand worker controlling robotic arms to complete a dangerous repair. Could BP's Deepwater Horizon environmental disaster of 2010 have been prevented with a 24/7 on-demand crew tasked with taking turns to monitor all industrial systems and repair them as sensors brought attention to any fitting starting to wear or needing to be replaced? Could companies prevent the next oil spill by approaching on-demand work not as contingent or expendable ghostly labor pools but as the collective intelligence of workers, controlling their schedules as they rotate as regulars who are on call, while others are always on standby — all equally valuable to their workplace?

The World Bank projects that the professional on-demand digital labor market, delivered through platforms like those we've studied, will grow to a $25 billion-a-year market by 2020. Regardless of whether we are ready to prioritize and support those doing ghost work, theirs represents the inevitable near- and (arguably) long-term future of jobs. As all service sector employment transforms, businesses, workers, and society are unlikely to face a swarm of robots taking over, but we will have to decide what we want ghost work to look like.[1]

Taken as a whole, the current "gig economy" — an ecosystem of independent contractors and small businesses driven by short-term projects that shift to meet market demands — is quietly moving to ghost work platforms. A growing number of people are picking up on-demand gig work online — accepting project-driven tasks, with companies that assign, schedule, route, and bill work through websites or mobile apps. It is difficult to gauge how many people do this work and how quickly it is growing. The United States does not routinely conduct a census of workers earning a living through what the Department of Labor calls "alternative work arrangements." The most recent census was published in 2017, after a 12-year gap. And, in the interest of maintaining historical continuity, the survey questions did not reference the internet. No one asked workers how digital technology might be, in part, managing their labor. Because people doing ghost work might show up in the survey under various headings, including "alternative work arrangements," "temporary staffing," and "self-employed," there is no easy way to tally the nation's on-demand labor force. The best estimate we have is from the 2016 Pew survey that found that 8 percent of U.S. working-age adults, roughly 20 million people, earned money doing tasks either offline or on.[2] That means approximately 12 out of every 100 working-age Americans already does some form of on-demand ghost work.

It is difficult to bring on-demand ghost work into full view until there is a more focused effort to measure it. Yet, statistically speaking, we need look no further than the subtle shifts happening in the current, so-called gig economy. This ecosystem of independent contractors and small businesses driven by short-term projects and resistance or incapacity to carry the costs of full-time employees is already quietly moving to on-demand work platforms. The rise of on-demand labor platforms signal the allure of using APIs to organize, route, and schedule contract work both to businesses and to people who need work. As the examples in this book suggest, merging contingent labor with new technologies fuels more than the AI revolution. It also takes over more mundane office-based "knowledge work," like online customer service and database entry. Collaborations between traditional staffing services, large companies, and platforms like Upwork reflect the next phase of development for on-demand employment.[3] A 2015 research study estimated that, in the U.S. and Europe alone, around 25 million people did some form of on-demand gig work online — accepting project-driven tasks from companies that assign, schedule, route, and bill work through websites or mobile apps.[4] If 25 million job opportunities seems small, consider that this type of job did not exist prior to the widespread adoption of web-based application programming interfaces (APIs) in the early 2000s. At this rate of growth, if combined with current trends in the growth of contract staffing and temp agency services, 60 percent of today's global employment will likely be converted into some form of ghost work by 2055.[5]

Currently, when an AI system that powers a phone app like Uber or a website like Facebook needs human help, it needs it fast. Traditional methods of hiring won't do, because end users won't tolerate too much delay in waiting for a car to arrive or a page to load. An always-on labor pool, accessible via an API, provides that necessary human input, on demand and at a moment's notice. Software developers can write code that automatically hires someone to do a task, checks their work, and pays them for their time and effort. On-demand ghost work platforms and their APIs allow humans to power many of the websites, apps, online services, and algorithms most consumers think are automated. Scientists and engineers will continue their quest toward full automation by trying to build an AI capable of human-level thought. Debating whether or not this is achievable and whether it would replace all workers distracts us from addressing the immediate needs of today's on-demand workforce. What is clear is that technologists will keep trying and, therefore, reproducing the paradox of automation's last mile. As the quest to building general-purpose AIs continues, human labor will be necessary, at least for the foreseeable future, to step in where current systems fall short.

Typically, when programmers sit down to write a line of code to execute a task, they use an interface defined by their computer's operating system to harness the power of their computer. But a CPU can only execute exactly those instructions it's given. So when programmers loop in on-demand labor to complete a task, they are also working with a person or groups of people sitting on the other side of an on-demand labor platform's API. The programmer now depends on a person performing on-demand ghost work to bring creativity to predictable code. Workers' ingenuity and availability finish the job designed by the programmer. Unlike computers, humans can both anticipate what they should do next and deliver a novel response, in a flash, to the unexpected. A person can deliberate, reflect on an expansive library of lived experiences, and confer with others to figure out what to do next to complete the job. In exchange for this boundless capacity to iterate and collaboratively remix solutions, humans bring something to work that CPUs cannot: ingenuity and novelty. This is the promise of on-demand work—the power of computation blended with the imagination and dynamism of human insight. So if companies want to capitalize on the promise of combining the power of computation with the creativity of human insight, they must care about the worker enough to ask: Who are the people who do on-demand work and what do they need, from technologies, businesses, consumers, and society, to work in concert with APIs?

We found three different profiles of on-demand workers mirroring the social dynamics of online communities described by a Pareto distribution. First, most workers try platforms out, do a few tasks, and leave within a week to a month. These workers offer value to the platform because they are counted as part of the potential labor market available for ghost work. Second, some workers regularly participate (for less than 10 hours a month) by picking up a set number of tasks, sometimes connected to a financial or time goal. Finally, some workers build up familiarity with a platform and become the core workforce, showing up daily and doing an average of 30 hours of work a week on the platform. These workers do the majority of the work available on the platform.

In all cases, APIs and on-demand labor platforms sit between workers and those hiring them, setting the terms of employment. When workers are not intentionally the center of an on-demand service's design, the platform strips away everything we now associate with a job, leaving behind the task, the pay, and the person doing the work. This can create sterile environments that offer no benefits, no advancement, no recognition, no leadership, no colleagues, no lunch buddies, and no after-work drinks. The APIs often go one step further, not only stripping away aspects of the work environment but also stripping away aspects of the workers and requesters themselves.

Different platforms do this to differing degrees but at one extreme each worker is identified by a sequence of seemingly random letters and numbers like "A16HE9ETNPNONN." These types of platforms remove all that makes each worker and requester human and reduces all involved to an identifier. There is no way to determine a worker's personal attributes such as gender, age, ambitions, or passions. Similarly, there is no way to understand the depth and breadth of a worker's skills, like how good they are at math, English, or drawing. Or how much better an individual is at any specific task when connected to a team. None of these questions have clear answers. For instance, an obvious downside to this approach is that it can lead to poor matches between workers and requesters.[6]

But an upside is that workers who may have been part of a discriminated class due to their gender or religion, for example, can now find work. At another extreme, there are platforms that allow workers to create profiles that list their skills and even include a photo. This might lead to better matches, but at the cost of potential discrimination and wage inequality. Overall, the anonymization and sometimes dehumanization of workers can make them seem interchangeable to requesters. For example, what's the difference between hiring A16HE9ETNPNONN and A6GQR3WXFSIYT?[7] Even if their attributes are not stripped away, it's easy to view workers as a pool to draw from, as opposed to individual people. And we have been here before.

## Learning from Ghost Work's Past

If history offers us any lessons for the future of work, it is that technologies do not eliminate the need for temporary labor. Technologies and our unquenchable thirst for experiencing the world produce job openings for a human's capacity to communicate, discern, and care about others. Our ability to reflect and update prior assumptions, without new information, and the verbal and nonverbal ways that we share feelings and thoughts with one another continue to be our edge, as humans rather than machines.[8]

Past generations of cultural traditions put full-time work on a pedestal, above all other types of employment. This made it difficult to see how much our collective intelligence and capacity to collaborate, more than any individual's professional title or social standing, made the difference in endeavors as small as keeping a spinning jenny running or as big as sending humans into space. As we saw in the case of the history of piecework to the current churn of tasks on today's on-demand platforms, contingent labor has always been necessary. But it's been systematically devalued, and its status as "less than" has been naturalized and codified into employment classifications.

Platform-driven innovations deliver goods and services to businesses and consumers under the pretense that a combination of APIs and artificial intelligence have eliminated what traditional employers used to pay for, namely, recruiting, training, and retaining workers. By spending time with hundreds of workers, we saw that automation, far from eliminating those costs, predominantly shifts them to workers. We offer guidance anchored in the successes and challenges of today's on-demand workers. We urge these workers to remember: *You are not the broken cog in this machine. You are doing everything that you can do to make this work. You can use your experience to teach us how to make the future of work dignified and more humane.*

The private companies investing in automation cannot afford to ignore what, beyond a paycheck, motivates people to work. APIs can distribute disaggregated tasks at scale and minimize the need for hands-on management. But these systems cannot eliminate a worker's desire to derive meaning from their working life. Rather than ignore or prohibit the human impulse to collaborate and connect, we show how the collaborative strategies of our research participants can inform software designs that accommodate—rather than ignore—the paradox of automation's last mile. In the short run, these strategies include developing systems of taskifying management itself. Companies could direct APIs to turn affirmation and encouragement into doable, paid tasks, compensating people for helping each other to do their best.

To understand how workers navigate the sometimes sterile on-demand environment, we circumvented the API. We interviewed workers in person, we studied how they go about their day, and we measured their activity in aggregate. Our key finding is that workers are putting the humanity and meaning back into their jobs. First and foremost, they are adding collaboration back into their work. They are re-creating the office watercooler, albeit in an online or virtual environment, to provide and receive social support. Workers clearly value peer collaboration—they are spending their own time and money building online forums, away from the platforms they work on, when they could be spending it working and earning. Workers also collaborate to find good work and reduce the overhead they experience in doing ghost work, which sheds light on some of the transaction costs that workers face. They don't get paid to search for tasks, to learn how to do them, or even how to learn how to use the on-demand labor platform in the first place. Ghost work shifts many of the transaction costs to the workers and requesters hiring them. So workers collaborate to reduce the toll of the unpaid work they have to do to make it in the on-demand economy.

Much of ghost work is automated—tasks like hiring, reviewing, and paying workers. But software is, by design, rigid. And its inability to make judgment calls results in three types of algorithmic cruelty. First, requesters can upload a large number of tasks at any time automatically using the API and disappear once they are done. Collaborating to find good work can help smooth out the bursty nature of when work arrives on the platform, helping workers offset the costs of the instability and insecurity that often come with this line of work. Nonetheless, collaboration does not fully solve the problem, as workers often feel the need to be always on call or hypervigilant to get good work before it's gone, which is the first form of algorithmic cruelty. Second, since the API governs the interaction between the worker and the requester, should the worker need guidance or assistance, they have no one to turn to. As a result, they often tap their work networks for help. Finally, platforms can unilaterally decide who has access and who doesn't, which implicitly means they decide who can earn and who can't. They use their own internal software tools to render a verdict on who stays or goes. Workers often have no avenue for recourse. These three forms of algorithmic cruelty are brought on by the ability to hire workers and give them access to a platform via APIs.

Workers persevered, in spite of these difficulties, because they could mold on-demand work to fit the constraints that they face. For example, cultural norms may dictate who can and can't work outside the home due to race, gender, religion, or disability status.[9] Since on-demand work can be done anonymously from anywhere, including from home, workers can use it to push against this barrier. Similarly, workers may have family obligations that limit their work hours. Since on-demand work can be available at any time, it can be molded around these responsibilities as well. Finally, if workers are constrained because they don't have the training for a job they seek, they can use on-demand work to build up a résumé of experience showing that they have what it takes to do a specific job.

## SEMI-AUTOMATED FUTURE

The days of large enterprises with full-time employees working on-site are numbered as more and more projects rely on an off-site workforce available on demand, around the globe. Our employment classification systems, won in the 1930s to make full-time assembly line work sustainable, were not built for this future.

As machines get more powerful and algorithms take over more and more problems, we know from past advances in natural language processing and image recognition that industries will continue to identify new problems to tackle. Thus, there is an ever-moving frontier between what machines can and can't solve. We call this the paradox of automation's last mile: as machines progress, the opportunity to automate something else appears on the horizon. This process constantly repeats, resulting in the expansion of automation through the perpetual creation and destruction of labor markets for new types of human labor. In other words, as machines solve more and more problems, we continue to identify needs for augmenting rather than replacing human effort. This paradox explains why on-demand services — commercial ventures that combine humans and APIs to source, schedule, manage, and deliver access to goods and services — are more likely to dominate the future of work than AI and bots alone.

## AUTOMATION VS. HUMAN LABOR IS A FALSE DICHOTOMY

As many in the world of AI and computing know, ghost work — the human computation behind augmented services — is a relatively new form of computing power that combines the speed and precision of software with human ingenuity. Ghost work is a great tool for solving problems that machines alone can't solve efficiently on their own. For example, MTurk workers generated training data used to power the algorithms that later solved easier pieces of the image recognition problem. Scientists and engineers then turned their attention to more sophisticated image-processing tasks. As AI progresses, we will see this pattern repeat using workers recruited from on-demand labor platforms to solve problems in new domains.

Even as ghost work has proven vital for the advancement of AI, it has transformed an informal crowd into the primary computing device, a distributed human computational system, for a range of on-demand services, from DoorDash and Upwork to Uber. These intelligent systems blend AI and humans in the loop to source, schedule, manage, and deliver access to a paid service through an API. They could be reworked to also value the kindness of strangers and recognize the strengths of online social networks as assets fundamental to good jobs in the future. As more people turn to ghost work — or have their formal employment turned into ghost work — we have a chance to learn from labor history and people's experiences of ghost work today to tackle its technical and social malfunctions. There is still time to bring jobs out of the shadow of AI and make them equitable and dignified employment of which all parties involved will feel proud. Drawing on what we have learned from research participants and studying the dynamics of on-demand labor markets, we have a few technical and social fixes to suggest for moving the future of work forward.

## What We Can Do Next to Update Jobs for the Future of Work

Imagine what society could achieve if businesses, consumers, governments, and citizens fully recognized the value of human creativity and the pooled, collective effort of so many humans contributing to on-demand services. What potential benefits could flow from projects channeled, at least in part, through APIs and software if we invested in them as a new category of employment, the most likely mode of employment through at least our grandchildren's lives?

To achieve the possibilities hinted at in the stories that opened this chapter — and avoid the disasters they might harbor — we first must recognize that on-demand platforms aren't just software. They are bustling, dynamic online labor markets that consist of *humans* on both sides of the market. These markets link an on-demand collective of workers to businesses that need human creativity and insights. They serve customers who enjoy the benefits of humans in the loop, whether they know it or not. The platforms facilitating the connection, when designed to recognize all humans involved as equally valuable and necessary to a business's bottom line, could open new work opportunities and reduce frictions that currently come with finding and hiring someone to start work on a particular job. Below we offer a list of our top recommendations for turning ghost work into sustainable, on-demand employment opportunities. As we hope the arguments in this book make clear, building better work is neither a social nor a technical challenge. It will always be both. Folding technology into our work lives requires the engineers and businesses building technologies to "think socially" about what people need in order to do good work; in equal measure, crafting the future of work requires that policy makers deeply understand the technologies introduced in the workplace. Each of the fixes below presses us to collectively blend our best technical and social expertise to design a better job future.

## Technical Fixes for Social Change

### FIX 1: COLLABORATION

Workers need APIs to facilitate rather than hinder communication among themselves. Just as workers controlling their schedules, projects, and connections to peers proved to strengthen their work, having greater ease in effectively communicating with one another and potential clients returns benefits. On-demand workers seek each other out. They build their own discussion forums. Yet many of today's on-demand services assume that those picking up tasks in this economy work independently and have no need to collaborate. Large projects are divided into seemingly simple, often repetitive tasks that one person could presumably do alone. Companies like LeadGenius use products like Google Docs and Voice over Internet Protocol (VoIP) conferencing software to harness workers' desire to collaborate. But most firms generating projects are currently missing the opportunity to leverage the capacity of human creativity to collaborate in this environment.

Our research clearly indicates that, despite the dismantling of full-time employment and on-site work environments generated by the shift to ghost work, humans value and thrive on the social nature of work. They put many unpaid hours into rebuilding social and professional connections with peers, and they prioritize opportunities that re-create long-term work. Short-term, API-mediated workflows cannot stop the human desire to connect through work. We propose "on-demand co-working spaces" that would allow multiple workers to simultaneously work together on the same shared document, spreadsheet, presentation, or program while video and text chatting. On-demand co-working spaces would also allow firms to generate shared workspaces for on-demand "flash teams" of individuals brought together to complete a project-specific task. This software would empower workers to work in teams on larger, more complex tasks using whatever organizational structure suits them.

## FIX 2: AN ON-DEMAND WATERCOOLER

On-demand ghost work environments can be isolating and lonely. Today's on-demand services can create sterile environments that strip meaningful employment down to a set of tasks and a paycheck. But humans are social animals who derive dignity and meaning from creating things together and developing professional relationships. Trying to prevent those doing ghost work from connecting is not only impossible but also misses the opportunity to capitalize on the value workers derive from connecting with one another. Our research indicates that workers who participate in online forums are far more likely to stay engaged and develop ties to on-demand services than those who do not. Today, people picking up ghost work turn to off-platform, community-built discussion boards, Skype, and social media like Facebook to start and break up their workday, root each other on, share information about potential work opportunities, and mentor one another in how to succeed in ghost work.

We propose a professional network to build guild-like communities for on-demand workers. These guild communities could foster and facilitate communication among those doing ghost work to share tips and tricks of the trade and to help individuals identify potential collaborators for future projects. Workers could share best practices, learn new skills, and get certified for those skills. In addition, an online guild community could serve as a portable reputation system that would document and showcase a worker's capacity and past performance. Requesters could reach out to members of these guilds to do their tasks, thereby increasing the likelihood that workers get matched to tasks that suit their skills. In the past, employers covered the costs of cultivating their employees' growth and advancement. This is now the distributed task of all those dipping in and out of this well of independent worker expertise and availability. Those stepping in and out of workflows need services and tools that deliver agency, transparency, and dignity rather than the stigma of being "between jobs." There is no more "between jobs." We are all preparing for our next task.

## FIX 3: COMPANY-ISSUED, SHARED WORKSPACE

The ability to contract on-demand labor, programmatically, will change knowledge work as we know it.

Companies are increasingly turning to on-demand ghost work platforms for deadline-driven projects, and this creates a continuous ebb and flow of workers across enterprise boundaries. Enterprises could assemble and equip combinations of full-time employees and individuals doing on-demand work to complete a firm's projects. Workers could then disperse within hours or minutes of accomplishing their goals. Again, to be able to do this, they need a shared workspace where everyone has access to the same tools and no one assumes that someone can—or should have to—absorb the costs of doing work that requires following a specific set of guidelines or procedures. When the investment in software is of greater benefit to the employer than to the worker, it's unfair to expect a worker to shoulder the costs.

Workers, contributing to bounded, time-sensitive projects within enterprise walls, need access to project resources that protect the security and privacy of all involved in the work. Right now, most ghost work platforms, particularly for micro-tasks working with AI training data, default to assuming that people are individual agents standing by and ready to jump into a task, with the latest software and a stable internet connection. For example, a manager might give a team of data scientists, hired through an online labor market, access to files and data, effectively bringing them into the enterprise for a short period of time. Then, once the project is completed, their access to the files and data would be revoked, returning them to a position outside of the enterprise. Managing the workflows—worker output and interactions with data—presents new challenges for porous enterprises mixing full-time employees and on-demand workers. We found people who never shared a time zone, let alone an organizational affiliation, bringing their own, often incompatible and out-of-date software and devices to a job. Current enterprise software typically builds in rigid enterprise boundaries, blocking or delaying off-site, on-demand collaborators. It doesn't have to be this way.

Companies need to shift their thinking and assume that they are relying on contributions and insights from people who won't always have the same access to updated software tools or the latest high-speed computer, or even a reliable electrical grid. In all cases, the value of on-demand work is contributed by the workers themselves. Assuming that providing their tools will improve the overall product is the first step toward rebuilding the business software tools necessary for an on-demand workplace. Work will be passed around over a network of independent and full-time workers, each one taking a chunk that best fits their capacities, solving their piece of the task while passing the rest on. All will be able to retrace the path of people who did work on any given task. But workers may not directly interact with the original requester and, similarly, the requesters initiating the work will not play a direct role in managing who will ultimately work on their tasks. It will become of increasingly paramount importance to keep track of who contributed which piece of intellectual property to the work and who received value from any given worker's contributions to the end product.

Building for this future is the top job for software developers around the world. Without shifting to an on-demand mindset, when it comes to business software, both companies and workers will be locked out of fully benefiting from the on-demand economy. Assume we are no longer "temporarily" including someone from outside the company on a project as the terms "inside" and "outside" the company will soon be an irrelevant distinction.

## FIX 4: ENABLE FLASH TEAMS FOR THE ENTIRE RANGE OF GHOST WORK, INCLUDING CONTENT MODERATION

Most on-demand work platforms could make one simple change that would make a world of difference for those workers trapped in a cycle of hypervigilance: create two clearly defined streams of on-demand work, one explicitly available for group collaboration and the other requiring independent, subjective results.

There are cases, such as generating training data for machine learning algorithms or survey responses, where independent results are required for validity. Where collaboration works against the desired outcomes, technical systems could focus on explicitly banning collaboration and telling workers why they cannot work together by building in more instructional transparency where the task quality depends on it. We could also focus our efforts on ferreting out breaches of the terms of participation where a worker's desire to collaborate really hurts the final outcome. Treating on-demand workers as potential security threats to be identified rather than workers to be trusted stokes the algorithmic cruelty that was such a common experience among the workers we met. Who wants to work in an environment where a manager follows them around, suspecting they will steal office supplies or turn in shoddy work at the drop of a hat?

There are many tasks, however, from sales lead generation to location verification, that do not, by design, require independent responses. When workers are expressly needed to collaborate on a project, it would be ideal if they could do so in what researchers call "flash teams"—rapidly generated teams, assembled for the course of a project, which then dissolve after project completion.[10] The findings on this approach to work are clear: workers' productivity goes up when they are able to work with others easily and have relatively loose ties imposed on them. When micro-task platforms assume that workers can—or should—work separately, they short-circuit the improvisational nature of the hallway check-ins and low-watt banter that keep people's energies up, even when they are short-term and contractual in nature.

Helping people connect—for example, by creating a shared "whiteboard" or one of the other communications options that LeadGenius and Amara use to structure their teams—creates better outcomes. And in cases where a requester really needs people to work without others' help, they should make that a feature of the work, rather than penalize workers who turn to one another to collaborate. The use case that seems like the most valuable to test today: creating content moderation flash teams that can support each other as they review online content for social media companies.

Whether it is fake news posts, hateful tweets or harmful speech spewed by vengeful trolls, or adult content that makes its way into YouTube channels, there is clearly a need to recognize the value of people curating and culling online content. As discussed throughout the book, recognizing and filtering out "bad content" is a technically hard problem. In some cases, online communities volunteer to moderate themselves. It's time for social media companies that benefit greatly from this volunteer work to pitch in to the commons and pay for the diligence of co-moderators willing to take on the trolls. This could look like, following Amara's example, the threading together of flash team members invested in content moderation as community work and, as the need arises, infusing those communities with the software upgrades, equipment, and pay to keep the commons of content moderation well nourished. There is no easy, free alternative, unless everyone decides to delete their social media accounts.

## FIX 5: RÉSUMÉ 2.0 AND PORTABLE REPUTATION SYSTEMS

Since requesters can seamlessly enter and exit the market, independent workers are often at a disadvantage when it comes to getting a rating or recommendation after they finish a task. On-demand workers will need reputation and recommendation systems that help them navigate finding their next income opportunity and manage the risk of such an uncertain future. They might establish a rapport with a requester for months, like Joan or Riyaz, only to find that requester leave the market or begin looking for workers with different skills. Successful workers will need to adapt to this dynamic environment quickly.

Systematically tallying and storing workers' accomplishments would be a boon to all involved. For example, a requester might hire a team of independent product managers and software architects and give them a product description; their task would be to write a software design. Once completed, this design would then be given to a new team of developers, perhaps with another product manager supporting them, and the output would be the code for the product. This code might then be given to a team of testers and usability experts for feedback. At any point along this process, there might be iteration between teams, and people may leave and return to teams. Work would flow into these teams, across enterprise boundaries, some of which would be passed on to others hired by teams coordinating specific pieces of the workflow. These teams would be solving problems that machines are not yet able to solve on their own. All play a role in pulling off the larger project and deserve credit they can use to land future work opportunities.

Many of the tools to implement the technical recommendations above are available today. The challenge is shifting our collective mindset to a new set of defaults regarding our dependency on—rather than the expendability of—on-demand labor. But, as discussed in chapter 6, companies hold back on sharing their tools, training, and other resources not because it's technically too hard to do but, rather, because they do not want to be seen as the "employer of record." The next set of recommendations take aim at making all of society the "employer of record" for on-demand work.

## Social Fixes That Demand Technical Expertise

### FIX 6: BUILD EMPATHY FOR WORKERS

One thing that readers can do is envision themselves or their family members someday working in the on-demand economy. Doing so will allow you to imagine walking a mile in a worker's shoes and confront the algorithmic cruelty they face simply for trying to earn a living or learn a new skill. But don't stop there. You could even try working on demand. At the very least, consider whether you are a customer of ghost work the next time you use a search engine or flag content online. Whatever role you choose, see that, like Joan and Asra, these actors behind the scenes are doing this work to provide for their loved ones. See that, just like Virginia and Gowri, they are building up their skills. See that they, too, like Justin and Kumuda, take pride in their work and pride in their contributions to their families.

By doing this, you'll see that on-demand labor platforms consist of humans, trying to make ends meet to support themselves and families, like all of us. You'll be able to see the humanity that the API has hidden. Allowing yourself to empathize with these workers will go a long way toward combating the "full-time work is better than part-time work" narrative we've heard for decades. The more you know about this community, the more you'll be able to effect change in it. And the more you participate, the quicker it will become clear that the social safety net needs to be overhauled to help on-demand workers. This future is still in its early days. It can be shaped. After all, you might start getting work through an API before you know it.

## FIX 7: INSTITUTE ACCOUNTABILITY AND AN ENTERPRISE-WIDE "GOOD WORK CODE" FOR GHOST WORK SUPPLY CHAINS

Currently, in an enterprise environment, any given task requires a decision, either implicit or explicit: Do we do the task "in-house" or hire out for that task? In large firms, or *enterprise* companies, the question may also be a matter of hiring talent from one part of the organization to draw on the capacities already in-house. In either case, with fewer hires needed for set, in-house tasks and more talented people necessary for rapid prototyping and testing new product ideas, many companies will have less need for full-time employees and an expanding dependency on an on-demand workforce at the ready for dynamic requests. While some transaction costs for hiring permanent workers will shrink, all businesses will continue to require some organization, at some cost, as Ronald Coase predicted.[11] The question is not whether the transaction costs disappear but, rather, who pays for them?

We need large corporations, which use vendor management systems to hire ghost work in bulk, to pull ghost work out of the shadows and into the daylight. As noted above, recent economic analyses of contingent labor markets confirm a sharp increase in both the amount and breadth of this type of employment.[12] Corporations using an on-demand approach for enterprise-level work have, so far, increased the fissuring of the workplace, driving up the number of self-employed, independent, and vendor-managed workers to meet business-to-business needs and consumer demands for ever-evolving AI-driven products.

As discussed, when a company's on-demand tools or platform crash or have technical difficulties, vendor-managed workers have to absorb the pain. A Good Work Code, modeled on the one developed by the National Domestic Workers Alliance, could push companies to make sure their vendors are there for people doing ghost work—or take their business to vendors who will.[13]

As on-demand workers fill more corporate staffing needs, often integrating several on-demand providers threaded together to complete different types of tasks, ghost work labor supply chains become harder to trace. Companies need to be able to guarantee to their customers and their own full-time employees that their products come from good work conditions and practices.

Companies could pledge, at minimum, to work with vendors who can verify that they provide a base pay for all work completed; pay workers within one week of completing a task; charge them no additional transfer fees to collect their paychecks; and have the same non-discrimination policies in place as the hiring company, to ensure that all workers producing value for a company are treated with equal dignity and respect. This code should also require companies to use vendors that can confirm, through audit, that they prohibit the use of child labor and that the tools they use to support workers are accessible to people with disabilities. Businesses that use online labor markets to hire people to complete tasks have a vested interest in making sure that the vendors hired to help the company accomplish its mission are using the same fair hiring practices that that company would use in-house. If hiring is going to become more like programming, where companies issue an open call for a task through an API and workers will compete to accept it, these companies must be part of leveling that competitive playing field. Tracking accountability along the on-demand labor supply chain will be integral to that mission.

## FIX 8: EMPLOYMENT CLASSIFICATION FIT FOR THE COMMONS

There is little doubt that the on-demand economy generates value and saves costs for businesses. In the process, it eliminates the traditional forms of stability and security attached to the overhead of full-time employment. The social upheaval that comes with such instability carries its own price. Is that the future that we want for ourselves and our loved ones? As the discussion of businesses applying a double bottom line suggests, this economy can grow when treated as a shared resource. Arguably, on-demand economies themselves will grow anemic and atrophy if they continue to be treated as expendable. These markets have their own business imperative for creating a sustainable labor commons rather than shortsightedly draining a temp pool. We will need to create job classifications that embrace the value of what is now called "the alternative work arrangement." Rather than think of anyone as working "full-time" or "part-time," let's shift to recognizing that people work when they can, as they can. No one is less valuable because they work "part-time." Besides, in an on-demand economy, this logic no longer makes sense.

The implications of a new employment classification might seem radical, but, in pragmatic ways, it is the necessary next step to catching up with a shift to a service economy that cannot do without contingency and rapid response to consumer demand. What would it look like to embrace the importance of contract work throughout all industries? Let's start with the basics and what would need to change to nurture rather than drain the commons.

## FIX 9: UNIONS AND PLATFORM COOPERATIVES AS THE NEW BETTER BUSINESS BUREAU

We need a third-party registry that allows on-demand workers to build their work résumés and accrue reputations, independent of the platform. Workers should be able to take their record of accomplishments with them, no matter where they pick up their next gig. Unlike a typical résumé, a registry, managed by workers' representatives, would allow workers to display validated feedback from previous employers and requesters, no matter the platform of origin. Platforms could be required to post this portable stack of letters of recommendation as part of the worker's platform-specific profile. These registries could operate as the on-demand economy's Better Business Bureau, authenticating workers' identities and reputations and saving companies the engineering costs currently poured into blocking the relatively few bad actors bombarding platforms with shoddy or fraudulent work. In exchange for bringing greater scrutiny to workers' reputations, registries could be used to hold companies accountable to workers' demands. For example, companies could be required to register suspensions and removals of workers.

Workers would then have a way to publicly appeal companies' actions when they feel that they have been unfairly blocked from work opportunities on a specific platform. Right now, workers have no formal rights to challenge account suspensions or the withholding of wages, leaving them with little or no recourse when they are blocked from using a specific platform to find work and given no route to appeal their case. We met several workers locked out of their platform accounts, with no prior warning and no clear explanation for why they had been removed from the site. Ideally, the U.S. Department of Labor could be part of this registry, so that it has a clearer way of supporting and keeping track of this rapidly expanding labor force often left uncounted. A registry could operate as a labor board for this world of contingent labor, which will need new strategies for collectively organizing and bargaining in this nebulous virtual workplace. Tasking labor organizations with safeguarding workers' reputations adds to their relevance for twenty-first-century work.

While traditional forms of labor organizing and union membership can make a difference, the future of workers' voices and bargaining power will have to contend with the absence of a shared physical work environment—making solidarity through co-presence a steeper hill to climb. It will also require that we observe how independent, on-demand workers already build collaborations and craft a sense of shared purpose even as they cultivate identities as self-employed contractors. We see that happening in our research, and it gives us hope.

## FIX 10: A SAFETY NET FOR FUTURE WORKERS

There are several key components to this safety net worth outlining in more detail.

### *Safety Net Part A: The on-demand business case for universal healthcare, paid family leave, municipal co-working space, and continuing education*

Recall the example of DoorDash, in the previous chapter, and how they connected their couriers to the third-party service Stride to find healthcare options through the Affordable Care Act. DoorDash, like its brick-and-mortar peer Walmart, currently relies on public subsidy of its workforce's healthcare needs. Workers like Joan and Zaffar have tenuous resources for contending with health needs. But if the future is dependent on workers who cannot turn to their employers of record to provide healthcare, society needs an alternative that reflects reality. Some make the case for universal healthcare, taking a charitable stance. The sick deserve our collective care. The cold bottom-line truth of on-demand economies is that companies need a healthy labor pool, and preventive, comprehensive care for all is the most cost-effective way to keep a pool of people healthy. And because being "between jobs" will no longer make sense, all citizens of adult working age should have access to paid family leave so that no one has to choose between caring for loved ones or a newborn and making a living.

Just as important is making sure that workers have healthy work conditions. The Occupational Safety and Health Administration (OSHA) could gain new relevance by equipping municipally sponsored co-working spaces with the ergonomics that will help people maintain their bodies while at work in front of a screen. A universal among always-on and many regulars is a home office space that allows them to step away from their workday. But if our interviews suggest anything about the future of work, it is that we are heading for a public health crisis brought on by people making ill-fitting chairs and folding tables suffice as a primary workspace. The cost savings of equipping libraries and other public institutions to formally serve the needs of the self-employed currently crowding local coffee shops would be welcome relief to those who need more space for work and more comfortable places to sit. Lastly, the outdated job retraining and résumé design courses that cost cities millions of dollars to run would be better spent if distributed as educational benefits to local residents so that they could attend courses at local, regional, and online institutions. Most of the demand in human intelligence will be for those able to puzzle through a problem and figure out what capacity they need to complete a task or find help to get the job done. Beyond the initial training and orientation to learning how to learn that comes with postsecondary college coursework, workers will need to constantly retool and rethink what educational steps to take next. A liberal arts education is a baseline necessity. Beyond that baseline, education is now part of "on-the-job" training and just as necessary for those hiring on-demand workers as the workers themselves.

### *Safety Net Part B: Retainer base wage for all working adults*

Some, like longtime union leader Andy Stern, of the Service Employees International Union (SEIU), argue that the time has come to turn to a universal basic income, or UBI.[14] This is not a new idea. It came into vogue with other Enlightenment ideals, like democracy. Early arguments for a basic income go like this: If citizens receive a basic income, the state can get out of the paternalistic business of managing the welfare state. It would no longer be in the role of deciding who deserves support or administering a system of doling out resources through a moral lens of who deserves help and what kind of help (cheese blocks vs. apples) would be most appropriate. We might understand this as an argument for the freedom of choice UBI could provide. A second argument, popularized by philosopher Philippe Van Parijs, suggests that nation-states should be invested in providing as much financial support as possible to all citizens to spend their time as they choose. Life should be about living and organizing societies, so Van Parijs argued. A country's coffers should be spent maximizing equal access to that opportunity.

The last way of thinking about the social value of UBI is as an asset to democracy itself. If someone cannot feed and house themselves, they can be coerced to support tyrannical rule to preserve their access to resources. Thomas Paine felt that UBI could guarantee the capacity of citizens to say no to despots hoping to exploit their economic desperation. UBI has come to the fore as a way to make sure workers with limited skills aren't left behind as technological progress races past them.

But universal basic income assumes that "low-skill" workers need an economic floor because they will have no options once machines beat them out for jobs. Here's what's wrong with that logic. This argument assumes that AI will eventually conquer all and people will be irrelevant to its advancement or extending the services that it proposes. But, as suggested by the history of task churn common to platforms, humans do not go away in the future of AI.

Human labor will be necessary for the foreseeable future. If we imagine AI and humans augmenting and supplementing each other, the next issue is not whether humans are necessary. The real question will be: When are they in demand, and for what purpose? The history of full-time employment has assumed that we need people available for set shifts in order to be most productive. On-demand information service economies operate on a different principle altogether. They assume that what you need most is good enough help, right now. To accomplish this, we need to support — to retain — a commons of on-demand workers, willing to make themselves available, on call. They are no longer assets of a single company to be invested in and held solely by that firm. They are needed to cover the myriad requests of consumers wherever and whenever there is demand for human creativity.

Proponents of UBI today, who put it forth as a salve for the wounds cut by automation, frame the conversation as "Those poor sops who are at the bottom of the skills ladder! Let's be charitable and give them a hand." This ignores how dependent the future of work will be on contract on-demand labor. Our response to the call for UBI is the economic case for companies and consumers to share in underwriting a retainer for anyone contributing to on-demand economies. Retainers, like those paid to lawyers and other professionals, acknowledge the collective need to have a healthy, available, continuously updating workforce to make on-demand economies sustainable. These new portable benefits would come from all adults of working age, government funds currently spent on administering public benefits, and corporate taxes collected and poured back into a commons pool to cover the social security benefits for retirement, retraining, and paid leave and health costs for each worker.[15] This workforce will require — and deserves — a different set of benefits and safety nets. The labor of hardworking people around the world should not be rendered invisible or opaque by the rise of AI. Other industries have already taught business the value of shoring up the more sustainable labor practices of their supply chains.

## A FIX FOR ALL OF US: CONSUMER ACTION

The Fair Food Program, implemented in 2011, took nearly 17 years of organizing and coalition building among students, workers, faith communities, and consumers across the United States. The FFP was started in Immokalee, Florida, by tomato workers. Through a lot of organizing effort, it expanded to tomato farms in Georgia, South Carolina, North Carolina, Maryland, Virginia, and New Jersey, as well as Florida strawberry and pepper farms. The initiative rallied large grocers and restaurateurs to play a role in lifting the wages and work conditions of some of the lowest-paid, most poorly treated farmworkers in the United States. McDonald's, Walmart, Trader Joe's, Sodexo, Aramark, Whole Foods, and Chipotle agreed to pay a small premium (1–4 cents per pound of produce) and pass it directly to farmworkers' payrolls. They also committed to buying produce only from growers who signed the Fair Food Code of Conduct, securing a range of improvements in work conditions. It is a case of what can be done collaboratively by consumers, workers, growers, and corporate buyers to produce better outcomes for business, the consumer, and those generating value for them both through contract, seasonal jobs.

Just as inspiring is the signing of Accord on Fire and Building Safety in Bangladesh, or the Bangladesh Accord. It followed the 2013 collapse of the Rana Plaza, in Bangladesh, the deadliest textile factory accident in history. The overcapacity facility, built with shoddy materials for nonindustrial use, with three illegal shop floors haphazardly tacked on to the building, crushed 1,100 workers to death and injured another 2,500. Five days later, protesters surrounded the central London Primark store to call out the company for being one of the hundreds of name-brand clothing manufacturers that had contributed to the negligence that resulted in the building's collapse and so many deaths. Months later, through the hard work of consumer boycotts and advocacy groups, in solidarity with unions representing Bangladeshi textile and garment workers, more than 200 brands agreed to acknowledge that because they profited from the labor these workers supplied and the value they created for their companies, they themselves bore responsibility for the safety of the workers in their supply chains. Companies like Aldi Nord, Aldi Süd, Primark, Puma, and American Eagle, all of whom used Rana Plaza to stitch together consumer goods, signed on. Even though some of the largest clothes makers using the plaza, like Walmart, Gap, Target, and Macy's, refused to sign the legally binding agreements to improve safety and conditions for workers, many of their European counterparts did sign. The Fair Food Program and the Bangladesh Accord are beacons for how to hold all of us accountable to the workers in manufacturing supply chains of the past. Just as consumers and labor played a critical role in putting pressure on food services and clothing manufacturers as they learned how their purchases put workers at risk, so can such awareness and efforts play a role in bringing sustainability and dignity to the on-demand economy.

And, just as we need companies to be accountable for the labor practices that produce our food and clothes, so, too, do we need them to be accountable to both consumers and workers producing and consuming on-demand tasks. Making the most of ghost work means learning this key fact: no one will build the killer app or platform for this kind of work future. The workers themselves — peer networks of people professionalizing and sharing professional knowledge — are truly the supercomputers here.

Like those people studied in this book, readers probably juggle multiple hats, as family members, workers, business owners, customers, and citizens swept up, knowingly or not, in ghost work. That means that we all have a role to play in the tasks at hand, for the technical and social fixes detailed above are not so easily divided among technologists, labor activists, and policy makers. They will require a collective effort. Bringing ghost work out of the shadows will require business owners, policy makers, customers, and citizens coming together. It will take our collective will to reorient employment toward a future that recognizes the valuable contributions we have to offer one another.

# ACKNOWLEDGMENTS

This book would not exist if scores of workers, engineers, and business leaders had not generously offered their time and, in several cases, opened their homes to share a bit of their lives with us. While we could profile only a handful of the people we met in this book, the strength of this project comes from every encounter with someone from this brave new world of work. We feel fortunate to have had the opportunity to learn from so many people. We thank you for trusting us with your insights and hope that we have represented your concerns and voices with as much fidelity and generosity as you extended to us.

And, as much as we are indebted to those willing to shed light on workers' perspectives of on-demand labor, their perspectives would've been impossible to gather and organize without the dedication and brilliance of our research team. We would like to thank Wei-Chu Chen, Sara Kingsley, Greg Minton, and Ming Yin for contributing to the original quantitative data analyses, and Syed Shoaib Ali, Deepti Kulkarni, Jason Qualls, and Kathryn Zyskowski for their contributions to the book's qualitative data analyses. Invaluable research assistance was provided along the way by Andrea Alarcón, Sarah Hamid, Rebecca Hoffman, Kate Miltner, Christopher Persaud, and Steven Schirra. Your commitment and intellectual insights are illustrative of the kindness of strangers and the power of collaboration that are at the core of this research.

## Mary's Acknowledgments

Over the course of writing this book, my spry, Depression-era-hardened parents weathered a series of health setbacks. Each trip out to see them made me appreciate the generous healthcare coverage and paid family leave programs provided by my employer. I benefited from the privilege of such care and couldn't have finished this book without it. Every time I interviewed an on-demand worker covering their family's entire healthcare bills, I got more angry about the gap between my employer-provided healthcare coverage and the dearth of options for those working on demand. In fact, the global on-demand economy depends on a distributed universe of workers and their families staying healthy. Corporations like mine have the most to gain from advocating that healthcare be made available to all who are or might someday contribute value to the on-demand economy. Hopefully this book offers a clear business case for ceasing the practice of leaving workers' health to luck.

I also greatly benefited from the generous intellectual brain trust and unprecedented collegiality that distinguish Microsoft Research. I want to start by thanking my coauthor, Siddharth Suri, for his willingness to embark on this uncharted exploration combining computer science and anthropology and staying the course, despite the potholes and squirrelly side trails along our way. Thanks also to collaborators Sarah Kingsley, Greg Minton, Jenn Wortman Vaughan, and Ming Yin, for your patience and willingness to take time away from your own research to teach me your "taste in questions" and approach to science. I am particularly grateful to two other members of our research team, Deepti and Shoaib, for making the fieldwork in India possible. You are exceptional ethnographers and scholars. I am glad to also count you as friends. Several arguments in this book would not exist without your diligence and fortitude.

My co-workers at the New England Lab were a constant source of generous feedback, encouragement, and patience as I experimented with breaking the boundaries of my already undisciplined scholarly training in anthropology, communication and media studies, and queer theory. And none of us would be able to exchange thoughts so freely and with such ease if not for the support of the exceptional Microsoft Research staff who keep us afloat. We, collectively, do our best work at MSR because of you.

My colleague Nancy Baym likes to say that the Microsoft Research Social Media Collective — the group that has become my intellectual universe — "collects amazing people." Our group's founder, danah boyd, collected me back in 2012 somewhat unexpectedly (at least to me), and I am so grateful to her for it. I want to thank the alumni and current members of the Social Media Collective orbiting Microsoft Research New England while I worked on this project: Ifeoma Ajunwa, Mike Ananny, Stacy Blasiola, Sarah Brayne, André Brock, Jed Brubaker, Jean Burgess, Robyn Caplan, Aleena Chia, Nick Couldry, Kate Crawford, Jade Davis, Judith Donath, Paul Dourish, Kevin Driscoll, Stefanie Duguay, Michaelanne Dye, Nicole Ellison, Megan Finn, Ysabel Gerrard, Kishonna Gray, Dan Greene, Germaine Halegoua, Caroline Jack, Henry Jenkins, Tero Karppi, Airi Lampinen, Jessa Lingel, Sonia Livingstone, Elena Maris, Annette Markham, Alice Marwick, J. Nathan Matias, Jolie Matthews, Shannon McGregor, Tressie McMillan Cottom, Josh McVeigh-Schultz, Andrés Monroy-Hernández, Sharif Mowlabocus, Dylan Mulvin, Laura Norén, Susanna Paasonen, Nick Seaver, Luke Stark, Jonathan Sterne, Tom Streeter, Jaroslav Švelch, Lana Swartz, TL Taylor, Katrin Tiidenberg, Penny Trieu, Siva Vaidhyanathan, and Shawn Walker. This amazing scholarly community, cultivated through the SMC, particularly its visiting researcher program, postdoctoral fellowship, and PhD internship, have shaped this book in immeasurable ways. Without exception, each SMC member models publicly engaged, critical scholarship that's lit a clear path for me to follow. The book's arguments, like all earnest creative work, comprise an amalgam of the SMC's generosity and support. You have nurtured my thinking throughout the research process, and I cannot say thank you enough. Suffice it to say, I will be buying each of you the first round of celebratory beverages for the foreseeable future. And

I owe a special thanks, not to mention an incalculable debt that all the cocktails in the world can never zero out, to Nancy Baym and Tarleton Gillespie. You make me happy to come to work and smarter every day. I could not have survived this expansive "stretch goal" of a project without you.

It took the leaders of the New England and New York City labs, Jennifer Chayes and Christian Borgs, to turn this project into a book. Thank you for trusting me with a lot of travel around the U.S. and to India and even more license to run with the academic freedom that you've baked into our lab. I wish every researcher had such unwavering support underwriting their interdisciplinary critical research.

I thank everyone at Houghton Mifflin Harcourt, especially Liz Anderson, Lori Glazer, Rosemary McGuinness, Michelle Triant, and Rick Wolff, for helping move our manuscript across the finish line, as well as Will Palmer for his deft copyediting. And I am grateful to Alison MacKeen, Jenny Stephens, and the team of experts at Sterling Lord Literistic, as well as Eamon Dolan, Joseph Calamia, Glenn Kramon, and other literary-minded people, for informing early iterations of the manuscript and putting it out into the world.

Fieldwork in Bangalore was made possible through the generous support of Microsoft Research India. Additional thanks to Danielle DeBroeck, Paul Estes, Eric Horvitz, Teresa Hutson, Jacki O'Neill, Igor Perisic, Nick Pyati, T. K. Rengarajan, and Jaime Teevan, who became fans and champions of our research at Microsoft. A special thanks goes to Rajesh Patel for his support and wisdom. Several chapters of the book were reworked through invited talks hosted by the following institutions: the AFL-CIO, the American Academy of Arts and Sciences, Economic Policy Institute, the Federal Reserve Board, Northwestern University, Oxford University, SEIU, Télécom ParisTech, University of California at San Diego, University of Massachusetts at Amherst, Université Paris-Sorbonne, and the University of Southern California. I also wish to thank Lauren Robel and my colleagues at Indiana University in the School of Informatics, Computing, and Engineering, the Media School, and the Anthropology and Gender Studies Departments for allowing me to create a bridge between IU and MSR. I am honored to be a part of the faculty. I hope that this book demonstrates the value of industry-based research labs and universities sharing the responsibility for studying the intersections of technology and society as well as training the next generation of scholars building socially accountable technologies.

To my friends in the Labor Tech Reading Group, particularly our fearless facilitator, Winifred Poster; friends in labor organizing, mindfulness, and tech who commented on specific pieces of the manuscript, particularly Aleli Alcala, Tom Alden, Lisa Braden-Harder, Antonio A. Casilli, Jessica Champagne, Nick Couldry, Steven Dawson, Paul Dourish, Natalie Foster, Walt Gantz, Lilly Irani, Dean Jansen, Henry Jenkins, Airi Lampinen, Rochelle LaPlante, Margaret Levi, Kristy Milland, Gina Neff, Tim O'Reilly, Shani Orgad, Anders Schneiderman, Julie Schor, Palak Shah, Rebecca Smith, Max Toth, and David Weil; and Harvard University's Berkman Klein Center for Internet and Society, particularly the thoughtful Yochai Benkler, Rebecca Tabasky, David Weinberger, and the fabulous Ethical Tech Working Group, with an extra shout-out to co-conspirator Kathy Pham. One and all read and/or suffered through listening to me talk through draft chapters. Thank you.

Lastly, to my extended chosen family, particularly T. L. Taylor, for your reads, our walks, and all the love and care you offered me, Andrew Brosnan, for coaching me through the darker mornings at my writing desk, and the Davis Women's Circle and their families, for reiterating that friendships are a powerful way to keep work in its place. You keep me grounded in reality and remind me to lighten up. And to the extended Guthrie and Tucker families for their love and relentless enthusiasm for this research. Thank you. I can't wait to join family activities without my laptop in tow.

Though I dedicated this book to my parents, Ila and George Tucker, in writing it I found a new depth of love and devotion for my partner in life and all things bookish, Catherine Guthrie. Neither of us realized that we would finish writing books within months of each other (whoops).

Witnessing the craft that Catherine applied to her book was humbling. Seeing people reimagine themselves through her carefully chosen words reminds me that sharing our experiences of the world matters. Catherine, you are my constant. Thank you for reading every sentence and helping me challenge my thinking and writing with compassion for myself and others. I cannot wait to see what we write next.

## Siddharth's Acknowledgments

First and foremost, I'd like to thank my wife, Miranda. Without her support, my part in this project would not have been possible. Thin or flush, we're a team, and I'm better for it. I'd also like to thank my mom and dad for always doing the right thing and for showing me the value of hard work. You were great parents and excellent role models. To Dan Goldstein, Jake Hofman, Winter Mason, and Sergei Vassilvitskii: you're all great scientists and even better friends. I'm grateful for all that I've learned from each of you and for our friendship. To my lifelong friends John Boyles, Marc and Stephanie Corliss, Boulos Harb, George Kanyi, Sean Knappenberger, Steve Muir, Jeb White, and Rick Zawadzki: I wouldn't have made it this far without each of you.

I've been fortunate to have a pioneering group of mentors throughout my career, including Max Mintz, Sampath Kannan, Michael Kearns, Jon Kleinberg, Éva Tardos, Duncan Watts, and Preston McAfee. The scientist I've managed to become is a result of your guidance. I thank Jennifer Chayes and Eric Horvitz for their "let 1,000 flowers bloom" approach to research. I hope you'll count this book as one of the blossoms.

I thank all the members of Microsoft Research, New York City, for making the lab such a lively and vibrant place to work. I owe a huge thank-you to all the interns, postdocs, and research assistants who contributed to this project including Shoaib Ali, Wei-Chu Chen, Sara Kingsley, Deepti Kulkarni, Greg Minton, and especially Ming Yin. I thank all of the workers who have participated in all of my research studies. Without you, none of this research would have been possible. Thanks to Nick Pyati and Danielle DeBroeck for their help in interviewing requesters. Special thanks to my coauthor, Mary, for asking me to join her on this project from the beginning. Finally, I thank Eamon Dolan, Rick Wolff, Rosemary McGuinness, Alison MacKeen, Jenny Stephens, and Will Palmer for their expert guidance in bringing our research into the public eye.

# BIBLIOGRAPHY

Abel, Marjorie, and Nancy Folbre. "A Methodology for Revising Estimates: Female Market Participation in the U.S. Before 1940." *Historical Methods: A Journal of Quantitative and Interdisciplinary History* 23, no. 4 (October 1, 1990): 167–76. https://doi.org/10.1080/01615440.1990.10594207.

Ahmed, Syed Ishtiaque, Nicola J. Bidwell, Himanshu Zade, Srihari H. Muralidhar, Anupama Dhareshwar, Baneen Karachiwala, Cedrick N. Tandong, and Jacki O'Neill. "Peer-to-Peer in the Workplace: A View from the Road." In *CHI '16: Proceedings of the 2016 CHI Conference on Human Factors in Computing Systems,* 5063–75. New York: ACM, 2016. https://doi.org/10.1145/2858036.2858393.

Albrecht, Sandra. "Industrial Home Work in the United States: Historical Dimensions and Contemporary Perspective." *Economic and Industrial Democracy* 3, no. 4 (1982): 414. https://doi.org/10.1177/0143831X8234003.

Amrute, Sareeta. *Encoding Race, Encoding Class: Indian IT Workers in Berlin.* Reprint. Durham, NC: Duke University Press, 2016.

Aneesh, A. *Neutral Accent: How Language, Labor, and Life Become Global.* Durham, NC: Duke University Press, 2015.

Ansel, Bridget, and Heather Boushey. *Modernizing U.S. Labor Standards for 21st-Century Families.* The Hamilton Project. Washington, DC: Brookings Institution, 2017.

Autor, David H. "Why Are There Still So Many Jobs? The History and Future of Workplace Automation." *Journal of Economic Perspectives* 29, no. 3 (Summer 2015): 3–30.

Baker, Dean. *Rigged: How Globalization and the Rules of the Modern Economy Were Structured to Make the Rich Richer.* Washington, DC: Center for Economic and Policy Research, 2016.

Barnard, John. *Walter Reuther and the Rise of the Auto Workers.* Boston: Little, Brown, 1983.

Barowy, Daniel W., Emery D. Berger, Daniel G. Goldstein, and Siddharth Suri. "VoxPL: Programming with the Wisdom of the Crowd." In *CHI '17: Proceedings of the 2017 CHI Conference on Human Factors in Computing Systems,* 2347–58. New York: ACM, 2017. https://doi.org/10.1145/3025453.

Barowy, Daniel W., Charlie Curtsinger, Emery D. Berger, and Andrew McGregor. "AutoMan: A Platform for Integrating Human-Based and Digital Computation." *Communications of the ACM* 59, no. 6 (June 2016): 102–109. https://doi.org/10.1145/2927928.

Basi, J. K. Tina. *Women, Identity and India's Call Centre Industry.* London: Routledge, 2009.

Battistoni, Alyssa. "The False Promise of Universal Basic Income." *Dissent,* Spring 2017. https://www.dissentmagazine.org/article/false-promise-universal-basic-income-andy-stern-ruger-bregman.

Board of Governors of the Federal Reserve System. *Report on the Economic Well-Being of U.S. Households in 2016.* Washington, DC: Federal Reserve Board, May 2017. https://www.federalreserve.gov/publications.htm.

Boudreau, Kevin J., Patrick Gaule, Karim R. Lakhani, Christoph Riedl, and Anita Williams Woolley. "From Crowds to Collaborators: Initiating Effort & Catalyzing Interactions Among Online Creative Workers." HBS Working Paper No. 14-060, Harvard Business School, Cambridge, MA, January 2014.

Boudreau, Kevin J., and Karim R. Lakhani. "Using the Crowd as an Innovation Partner." *Harvard Business Review* 91, no. 4 (2013): 60–69.

Bowe, John, Marisa Bowe, and Sabin Streeter, eds. *Gig: Americans Talk About Their Jobs.* New York: Broadway Books, 2001.

Brenner, Chris. *Work in the New Economy: Flexible Labor Markets in Silicon Valley.* Malden, MA: Wiley-Blackwell, 2002.

Broadband Commission for Sustainable Development. *State of Broadband 2017: Broadband Catalyzing Sustainable Development.* Geneva, Switzerland: Broadband Commission for Sustainable Development, 2017.

Brynjolfsson, Erik, and Andrew McAfee. *Race Against the Machine: How the Digital Revolution Is Accelerating Innovation, Driving Productivity, and Irreversibly Transforming Employment and the Economy.* Lexington, MA: Digital Frontier, 2012.

——. *The Second Machine Age: Work, Progress, and Prosperity in a Time of Brilliant Technologies.* New York: W. W. Norton, 2014.

Bureau of Labor Statistics. "Contingent and Alternative Employment Arrangements, May 2017." Economic News Release, U.S. Department of Labor, June 7, 2018.

Butler, Elizabeth Beardsley. "Women and the Trades: Pittsburgh, 1907–1908." Pittsburgh: Charities Publication Committee, 1909.

Chandler, Jesse, Pam Mueller, and Gabriele Paolacci. "Nonnaïveté Among Amazon Mechanical Turk Workers: Consequences and Solutions for Behavioral Researchers." *Behavior Research Methods* 46, no. 1 (March 2014): 112–30, https://doi.org/10.3758/s13428-013-0365-7.

Chen, Julie Yujie. "Thrown Under the Bus and Outrunning It! The Logic of Didi and Taxi Drivers' Labour and Activism in the On-Demand Economy." *New Media & Society* 20, no. 8 (September 6, 2017): 2691–711. https://doi.org/10.1177/1461444817729149.

Chen, Wei-Chu, Mary L. Gray, and Siddharth Suri. "More than Money: Correlation

Among Worker Demographics, Motivations, and Participation in Online Labor Markets." Under review, ICWSM '19: The 13th International AAAI Conference on Web and Social Media, Munich, Germany, June 2019.

Clawson, Dan, and Naomi Gerstel. *Unequal Time: Gender, Class, and Family in Employment Schedules.* New York: Russell Sage Foundation, 2014.

Coase, R. H. "The Nature of the Firm." *Economica* 4, no. 16 (1937): 386–405. https://doi.org/10.1111/j.1468-0335.1937.tb00002.x.

Coca, Nithin. "Nurses Join Forces with Labor Union to Launch Healthcare Platform Cooperative." Shareable. Accessed June 21, 2018. https://www.shareable.net/blog/nurses-join-forces-with-labor-union-to-launch-healthcare-platform-cooperative.

"Common Ground for Independent Workers." *From the WTF? Economy to the Next Economy* (blog), November 10, 2015. https://wtfeconomy.com/common-ground-for-independent-workers-83f3fbcf548f#.ey89fvtnn.

Cowan, Ruth Schwartz. *More Work for Mother: The Ironies of Household Technology from the Open Hearth to the Microwave.* 2nd ed. New York: Basic Books, 1985.

Daso, Frederick. "Bill Gates and Elon Musk Are Worried for Automation — But This Robotics Company Founder Embraces It." *Forbes,* December 18, 2017. https://www.forbes.com/sites/frederickdaso/2017/12/18/bill-gates-elon-musk-are-worried-about-automation-but-this-robotics-company-founder-embraces-it/.

Dayton, Eldorous. *Walter Reuther: The Autocrat of the Bargaining Table.* New York: Devin-Adain, 1958.

Deng, J., W. Dong, R. Socher, L. Li, Kai Li, and Li Fei-Fei. "ImageNet: A Large-Scale Hierarchical Image Database." In *2009 IEEE Conference on Computer Vision and Pattern Recognition,* 248–55. Piscataway, NJ: IEEE. https://doi.org/10.1109/CVPR.2009.5206848.

Denyer, Simon. *Rogue Elephant: Harnessing the Power of India's Unruly Democracy.* New York: Bloomsbury Press, 2014.

DePillis, Lydia. "The Next Labor Fight Is Over When You Work, Not How Much You Make." *Wonkblog* (blog), *Washington Post,* May 8, 2015. https://www.washingtonpost.com/news/wonk/wp/2015/05/08/the-next-labor-fight-is-over-when-you-work-not-how-much-you-make.

Difallah, Djellel, Elena Filatova, and Panos Ipeirotis. "Demographics and Dynamics of Mechanical Turk Workers." In *Proceedings of the Eleventh ACM International Conference on Web Search and Data Mining,* 135–43. New York: ACM Press, 2018. https://doi.org/10.1145/3159652.3159661.

Downey, Greg. "Virtual Webs, Physical Technologies, and Hidden Workers." *Technology and Culture* 42, no. 2 (April 2001): 209–35.

Dube, Arindrajit, Jeff Jacobs, Suresh Naidu, and Siddharth Suri. "Monopsony in Online Labor Markets." *American Economic Review: Insights,* forthcoming.

Duffy, Brooke Erin. *(Not) Getting Paid to Do What You Love: Gender, Social Media, and Aspirational Work.* New Haven, CT: Yale University Press, 2017.

Ekbia, H. R., and Bonnie A. Nardi. *Heteromation, and Other Stories of Computing and Capitalism*. Cambridge, MA: MIT Press, 2017.

Erickson, Kristofer, and Inge Sørensen. "Regulating the Sharing Economy." *Internet Policy Review* 5, no. 3 (June 30, 2016). https://doi.org/10.14763/2016.2.414.

Farrell, Diana, and Fiona Greig. *The Online Platform Economy: Has Growth Peaked?* JPMorgan Chase Institute, 2017.

Folbre, Nancy. "Women's Informal Market Work in Massachusetts, 1875–1920." *Social Science History* 17, no. 1 (1993): 135–60. https://doi.org/10.2307/1171247.

Foroohar, Rana. "We're About to Live in a World of Economic Hunger Games." *Time*, July 19, 2016. http://time.com/4412410/andy-stern-universal-basic-income/.

Fort, Karën, Gilles Adda, and K. Bretonnel Cohen. "Amazon Mechanical Turk: Gold Mine or Coal Mine?" *Computational Linguistics* 37, no. 2 (2011): 413–20.

Foster, John Bellamy, Robert W. McChesney, and R. Jamil Jonna. "The Global Reserve Army of Labor and the New Imperialism." *Monthly Review* 63, no. 6 (2011): 1.

Frahm, Jill. "The Hello Girls: Women Telephone Operators with the American Expeditionary Forces During World War I." *Journal of the Gilded Age and Progressive Era* 3, no. 3 (2004): 271–93.

Gabler, Neal. "The Secret Shame of Middle-Class Americans." *The Atlantic*, May 2016. https://www.theatlantic.com/magazine/archive/2016/05/my-secret-shame/476415/.

Gershon, Ilana. *Down and Out in the New Economy: How People Find (or Don't Find) Work Today*. Chicago: University of Chicago Press, 2017.

Gillespie, Tarleton. *Custodians of the Internet: Platforms, Content Moderation, and the Hidden Decisions That Shape Social Media*. New Haven, CT: Yale University Press, 2018.

——. "The Politics of 'Platforms.'" *New Media & Society* 12, no. 3 (May 1, 2010): 347–64. https://doi.org/10.1177/1461444809342738.

Graham, Mark, Isis Hjorth, and Vili Lehdonvirta. "Digital Labour and Development: Impacts of Global Digital Labour Platforms and the Gig Economy on Worker Livelihoods." *Transfer: European Review of Labour and Research* 23, no. 2 (2017): 135–62. https://doi.org/10.1177/1024258916687250.

Gray, Mary L., Siddharth Suri, Syed Shoaib Ali, and Deepti Kulkarni. "The Crowd Is a Collaborative Network." In *CSCW '16. Proceedings of the 19th ACM Conference on Computer-Supported Cooperative Work & Social Computing*, 134–47. New York: ACM, 2016. https://doi.org/10.1145/2818048.2819942.

Green, Francis. *Skills and Skilled Work: An Economic and Social Analysis*. Oxford, England: Oxford University Press, 2013.

Greenhouse, S. "Equal Work, Less-Equal Perks: Microsoft Leads the Way in Filling Jobs with 'Permatemps.'" *New York Times*, March 30, 1998.

Gregg, Melissa. *Work's Intimacy*. Cambridge, England: Polity, 2011.

——. *Counterproductive: Time Management in the Knowledge Economy*. Durham, NC: Duke University Press, 2018.

Grier, David Allen. *When Computers Were Human*. Princeton, NJ: Princeton University Press, 2005.

Grossman, Jonathan. "Fair Labor Standards Act of 1938: Maximum Struggle for a Minimum Wage." Office of the Assistant Secretary for Administration and Management, United States Department of Labor website. Originally published in *Monthly Labor Review,* June 1978. https://www.dol.gov/oasam/programs/history/flsa1938.htm.

Hamari, Juho, Mimmi Sjöklint, and Antti Ukkonen. "The Sharing Economy: Why People Participate in Collaborative Consumption." *Journal of the Association for Information Science and Technology,* 2015.

Hamill, Jasper. "Elon Musk's Fears of AI Destroying Humanity Are 'Speciesist', Said Google Boss." *Metro,* May 2, 2018. https://metro.co.uk/2018/05/02/elon-musks-fears-artificial-intelligence-will-destroy-humanity-speciesist-according-google-founder-larry-page-7515207/.

Hara, Kotaro, Abi Adams, Kristy Milland, Saiph Savage, Chris Callison-Burch, and Jeffrey Bigham. "A Data-Driven Analysis of Workers' Earnings on Amazon Mechanical Turk." 2018 CHI Conference on Human Factors in Computing Systems, Paper No. 449, 2018.

Harhoff, Dietmar, and Karim R. Lakhani, eds. *Revolutionizing Innovation: Users, Communities, and Open Innovation*. Cambridge, MA: MIT Press, 2016.

Harris, LaShawn. *Sex Workers, Psychics, and Numbers Runners: Black Women in New York City's Underground Economy*. Urbana: University of Illinois Press, 2016.

Hartley, Scott. *The Fuzzy and the Techie: Why the Liberal Arts Will Rule the Digital World*. Boston: Houghton Mifflin Harcourt, 2017.

Hatton, Erin. *The Temp Economy: From Kelly Girls to Permatemps in Postwar America*. Philadelphia: Temple University Press, 2011.

Hawksworth, John, Barret Kupelian, Richard Berriman, and Duncan Mckellar. *UK Economic Outlook: Prospects for the Housing Market and the Impact of AI on Jobs*. London: PricewaterhouseCoopers, 2017.

Herzenberg, Stephen A., John A. Alic, and Howard Wial. *New Rules for a New Economy: Employment and Opportunity in Post-Industrial America*. Ithaca, NY: ILR Press, 2000.

Hill, Steven. *How (Not) to Regulate Disruptive Business Models*. Berlin: Friedrich Ebert Stiftung, 2016.

Hochschild, Arlie Russell. *The Managed Heart: Commercialization of Human Feeling*. 3rd ed. Berkeley: University of California Press, 2012.

Hochschild, Arlie, and Anne Machung. *The Second Shift: Working Families and the Revolution at Home*. Rev. ed. New York: Penguin, 2012.

Holt, Nathalia. *Rise of the Rocket Girls: The Women Who Propelled Us, from Missiles to the Moon to Mars*. Reprint. New York: Back Bay, 2017.

Horowitz, Sara. "Special Report: The Costs of Nonpayment." *Freelancers Union Blog,* accessed May 8, 2018. http://blog.freelancersunion.org/2015/12/10/costs-nonpayment/.

Horton, John. "Online Labor Markets." In *Internet and Network Economics: 6th International Workshop, WINE 2010, Stanford, CA, USA, December 13–17, 2010, Proceedings,* Lecture Notes in Computer Science New York. Springer, 2011.

Humphreys, Lee. *The Qualified Self: Social Media and the Accounting of Everyday Life.* Cambridge, MA: MIT Press, 2018.

Huws, Ursula. *Labor in the Global Digital Economy: The Cybertariat Comes of Age.* New York: Monthly Review Press, 2014.

Hyman, Louis. *Temp: How American Work, American Business, and the American Dream Became Temporary.* New York: Penguin, 2018.

Ipeirotis, Panagiotis G. "Analyzing the Amazon Mechanical Turk Marketplace." *XRDS: Crossroads, The ACM Magazine for Students* 17, no. 2 (December 1, 2010): 16. https://doi.org/10.1145/1869086.1869094.

Ipeirotis, Panos. "How Many Mechanical Turk Workers Are There?" *A Computer Scientist in Business School* (blog), January 29, 2018. http://www.behind-the-enemy-lines.com/.

Irani, Lilly C., and M. Six Silberman. "Turkopticon: Interrupting Worker Invisibility in Amazon Mechanical Turk." In *CHI '13: Proceedings of the SIGCHI Conference on Human Factors in Computing Systems.* New York: ACM, 2013.

Isaac, Mike, and Noam Scheiber. "Uber Settles Cases with Concessions, but Drivers Stay Freelancers." *New York Times,* April 21, 2016. http://www.nytimes.com/2016/04/22/technology/uber-settles-cases-with-concessions-but-drivers-stay-freelancers.html.

Jarrett, Kylie. *Feminism, Labour and Digital Media: The Digital Housewife.* New York: Routledge, 2015.

Katz, Lawrence, and Alan Krueger. "The Rise and Nature of Alternative Work Arrangements in the United States, 1995–2015." NBER Working Paper Series, no. 22667, National Bureau of Economic Research, Cambridge, MA, September 2016. https://doi.org/10.3386/w22667.

Kingsley, Sara Constance, Mary L. Gray, and Siddharth Suri. "Accounting for Market Frictions and Power Asymmetries in Online Labor Markets." *Policy & Internet* 7, no. 4 (December 1, 2015): 383–400. https://doi.org/10.1002/poi3.111.

Kuehn, Kathleen, and Thomas F. Corrigan. "Hope Labor: The Role of Employment Prospects in Online Social Production." *Political Economy of Communication* 1, no. 1 (May 16, 2013). http://www.polecom.org/index.php/polecom/article/view/9.

Kuek, Siou Chew, Cecilia Paradi-Guilford, Toks Fayomi, Saori Imaizumi, Panos Ipeirotis, Patricia Pina, and Manpreet Singh. *The Global Opportunity in Online Outsourcing.* Washington, DC: World Bank Group, June 2015.

Kulkarni, Anand, Philipp Gutheim, Prayag Narula, David Rolnitzky, Tapan Parikh, and Bjorn Hartmann. "MobileWorks: Designing for Quality in a Managed Crowdsourcing Architecture." *IEEE Internet Computing* 16, no. 5 (September 2012): 28–35. https://doi.org/10.1109/MIC.2012.72.

Lambert, Susan J., Peter J. Fugiel, and Julia R. Henly. "Precarious Work Schedules

Among Early-Career Employees in the US: A National Snapshot." Research brief, EINet at the University of Chicago, 2014.

Lampinen, Airi, Victoria Bellotti, Andrés Monroy-Hernández, Coye Cheshire, and Alexandra Samuel. "Studying the 'Sharing Economy': Perspectives to Peer-to-Peer Exchange." In *CSCW '15 Companion: Proceedings of the 18th ACM Conference Companion on Computer Supported Cooperative Work & Social Computing*, 117–21. New York: ACM, 2015. https://doi.org/10.1145/2685553.2699339.

Leopold, Till Alexander, Saadia Zahidi, and Vesselina Ratcheva. *The Future of Jobs: Employment, Skills and Workforce Strategy for the Fourth Industrial Revolution.* Geneva, Switzerland: World Economic Forum, 2016.

Levy, Frank, and Richard Murnane. *The New Division of Labor: How Computers Are Creating the Next Job Market*. Princeton, NJ: Princeton University Press, 2004.

Li, Fei-Fei. "ImageNet: Where Have We Been? Where Are We Going?" ACM Learning Webinar, 2017. https://learning.acm.org/.

Lichtenstein, Nelson. *The Most Dangerous Man in Detroit*. New York: Basic Books, 1995.

Light, Jennifer. "When Computers Were Women." *Technology and Culture* 40, no. 3 (July 1999): 455–83.

Manyika, James, Michael Chui, Mehdi Miremadi, Jacques Bughin, Katy George, Paul Willmott, and Martin Dewhurst. *Harnessing Automation for a Future That Works.* Washington, DC: McKinsey Global Institute, January 2017. http://www.mckinsey.com/global-themes/digital-disruption/harnessing-automation-for-a-future-that-works.

Manyika, James, Susan Lund, Jacques Bughin, Kelsey Robinson, Jan Mischke, and Deepa Mahajan. *Independent Work: Choice, Necessity, and the Gig Economy.* Washington, DC: McKinsey Global Institute, October 2016. http://www.mckinsey.com/global-themes/employment-and-growth/independent-work-choice-necessity-and-the-gig-economy.

Marwick, Alice E. *Status Update: Celebrity, Publicity, and Branding in the Social Media Age.* New Haven, CT: Yale University Press, 2013.

Mason, Winter, and Siddharth Suri. "Conducting Behavioral Research on Amazon's Mechanical Turk." *Behavior Research Methods* 44, no. 1 (March 2012): 1–23. https://doi.org/10.3758/s13428-011-0124-6.

Meyer, Eric. "Inadvertent Algorithmic Cruelty." *Meyerweb* (blog), December 24, 2014. https://meyerweb.com/eric/thoughts/2014/12/24/inadvertent-algorithmic-cruelty/; revised version published on Slate.com, December 29, 2014.

"Middle Skills — U.S. Competitiveness." Harvard Business School. Accessed May 22, 2018. https://www.hbs.edu/competitiveness/research/Pages/middle-skills.aspx.

Mishel, Lawrence. *Uber and the Labor Market*. Washington, DC: Economic Policy Institute, 2018. https://www.epi.org/publication/uber-and-the-labor-market-uber-drivers-compensation-wages-and-the-scale-of-uber-and-the-gig-economy/.

Mishel, Lawrence, Elise Gould, and Josh Bivens. *Wage Stagnation in Nine Charts.*

Washington, DC: Economic Policy Institute, 2015. http://www.epi.org/publication/charting-wage-stagnation/.

Murphy, Kevin P. *Machine Learning: A Probabilistic Perspective.* Cambridge, MA: MIT Press, 2012.

Nadeem, Shehzad. *Dead Ringers: How Outsourcing Is Changing the Way Indians Understand Themselves.* Reprint. Princeton, NJ: Princeton University Press, 2013.

Nantz, Jay Shambaugh, Ryan Nunn, Patrick Liu, and Greg Nantz. *Thirteen Facts About Wage Growth.* Washington, DC: Brookings Institution, September 25, 2017. https://www.brookings.edu/research/thirteen-facts-about-wage-growth/.

*National Labor Relations Board v. Hearst Publications.* 322 U.S. 111 (1944).

National Low Income Housing Coalition. "Out of Reach." Washington, DC: National Low Income Housing Coalition, 2018. http://nlihc.org/oor.

Neff, Gina. *Venture Labor: Work and the Burden of Risk in Innovative Industries.* Cambridge, MA: MIT Press, 2012.

Newitz, Annalee. "The Secret Lives of Google Raters." *Ars Technica*, April 27, 2017. https://arstechnica.com/features/2017/04/the-secret-lives-of-google-raters/.

Noble, David. *Forces of Production: A Social History of Industrial Automation.* New York: Routledge, 2017.

Noble, Safiya Umoja, and Brendesha M. Tynes, eds. *The Intersectional Internet: Race, Sex, Class and Culture Online.* New York: Peter Lang, 2016.

O'Neil, Cathy. *Weapons of Math Destruction: How Big Data Increases Inequality and Threatens Democracy.* New York: Crown, 2016.

Painter, Nell Irvin. *The History of White People.* Reprint. New York: W. W. Norton, 2011.

——. "What Is Whiteness?" *New York Times,* December 21, 2017. https://www.nytimes.com/2015/06/21/opinion/sunday/what-is-whiteness.html.

Pal, Mahuya, and Patrice Buzzanell. "The Indian Call Center Experience: A Case Study in Changing Discourses of Identity, Identification, and Career in a Global Context." *Journal of Business Communication* 45, no. 1 (January 1, 2008): 31–60. https://www.doi.org/10.1177/0021943607309348.

Parry, Thomas Fox. "The Death of a Gig Worker." *The Atlantic,* June 1, 2018. https://www.theatlantic.com/technology/archive/2018/06/gig-economy-death/561302/?utm_source=atltw.

Pennington, Shelley, and Belinda Westover. "Types of Homework." In *A Hidden Workforce,* 44–65. Women in Society series. London: Palgrave Macmillan, 1989. https://doi.org/10.1007/978-1-349-19854-2_4.

Perez, Tom E., and Penny Pritzker. "A Joint Imperative to Strengthen Skills." *The Commerce Blog.* U.S. Department of Commerce website, September 11, 2013.

Piketty, Thomas. *Capital in the Twenty-First Century.* Translated by Arthur Goldhammer. Reprint. Cambridge, MA: Belknap Press of Harvard University Press, 2017.

Poster, Winifred R. "Hidden Sides of the Credit Economy: Emotions, Outsourcing, and Indian Call Centers." *International Journal of Comparative Sociology* 54, no. 3 (June 2013): 205–27. https://doi.org/10.1177/0020715213501823.

Prassl, Jeremias. *Humans as a Service: The Promise and Perils of Work in the Gig Economy.* Oxford, England: Oxford University Press, 2018.

Raghuram, Sumita. "Identities on Call: Impact of Impression Management on Indian Call Center Agents." *Human Relations* 66, no. 11 (November 1, 2013): 1471–96. http://doi.org/10.1177/0018726713481069.

Reich, Robert. "How the New Flexible Economy Is Making Workers' Lives Hell." *Robert Reich* (blog), April 20, 2015. http://robertreich.org/post/116924386855.

Reuther, Victor. *The Brothers Reuther and the Story of the UAW.* Boston: Houghton Mifflin, 1976.

Roberts, Sarah T. *Behind the Screen: Content Moderation in the Shadows of Social Media.* New Haven, CT: Yale University Press, forthcoming.

——. "Digital Detritus: 'Error' and the Logic of Opacity in Social Media Content Moderation." *First Monday* 23, no. 3 (March 1, 2018). http://firstmonday.org/ojs/index.php/fm/article/view/8283.

——. "Social Media's Silent Filter." *The Atlantic,* March 8, 2017. https://www.theatlantic.com/technology/archive/2017/03/commercial-content-moderation/518796/.

Roediger, David R. *The Wages of Whiteness: Race and the Making of the American Working Class.* London: Verso, 1999.

Rosenblat, Alex. *Uberland: How Algorithms Are Rewriting the Rules of Work.* Oakland: University of California Press, 2018.

Rosenblat, Alex, and Luke Stark. "Algorithmic Labor and Information Asymmetries: A Case Study of Uber's Drivers." *International Journal of Communication* 10 (July 27, 2016): 27.

Ross, Alex. *The Industries of the Future.* New York: Simon and Schuster, 2016.

Rustrum, Chelsea. "Q&A with Felix Weth of Fairmondo, the Platform Co-Op That's Taking on eBay." Shareable. Accessed June 21, 2018. https://www.shareable.net/blog/qa-with-felix-weth-of-fairmondo-the-platform-co-op-thats-taking-on-ebay.

Salehi, Niloufar, Lilly C. Irani, Michael S. Bernstein, Ali Alkhatib, Eva Ogbe, Kristy Milland, and Clickhappier. "We Are Dynamo: Overcoming Stalling and Friction in Collective Action for Crowd Workers." In *CHI '15: Proceedings of the 33rd Annual ACM Conference on Human Factors in Computing Systems,* 1621–30. New York: ACM, 2015. http://doi.org/10.1145/2702123.2702508.

Scholz, Trebor. *Uberworked and Underpaid: How Workers Are Disrupting the Digital Economy.* Cambridge, England: Polity, 2016.

Schor, Juliet B. "Debating the Sharing Economy." Great Transition Initiative, October 2014. http://www.greattransition.org/publication/debating-the-sharing-economy.

Schor, Juliet B., Connor Fitzmaurice, Lindsey B. Carfagna, Will Attwood-Charles, and Emilie Dubois Poteat. "Paradoxes of Openness and Distinction in the Sharing Economy." *Poetics* 54 (2016): 66–81.

Schor, Juliet B., and Craig J. Thompson. *Sustainable Lifestyles and the Quest for Plenitude: Case Studies of the New Economy.* New Haven, CT: Yale University Press, 2014.

Schuman, M. "History of Child Labor in the United States—Part 1: Little Children Working." *Monthly Labor Review,* U.S. Bureau of Labor Statistics, January 2017. https://www.bls.gov/opub/mlr/2017/article/history-of-child-labor-in-the-united-states-part-1.htm.

Schwab, Klaus. *The Fourth Industrial Revolution.* New York: Penguin, 2017.

Sharma, Dinesh C. *The Outsourcer: The Story of India's IT Revolution.* Cambridge, MA: MIT Press, 2015.

Shestakofsky, Benjamin. "Working Algorithms: Software Automation and the Future of Work." *Work and Occupations,* August 28, 2017. https://doi.org/10.1177/0730888417726119.

Shetterly, Margot Lee. *Hidden Figures: The American Dream and the Untold Story of the Black Women Mathematicians Who Helped Win the Space Race.* Media tie-in edition. New York: William Morrow, 2016.

Silberman, M. Six. "Human-Centered Computing and the Future of Work: Lessons from Mechanical Turk and Turkopticon, 2008–2015." PhD diss., University of California, Irvine, 2015.

Silberman, M. Six, and Lilly Irani. "Operating an Employer Reputation System: Lessons from Turkopticon, 2008–2015." *Comparative Labor Law & Policy Journal,* February 8, 2016.

Silver, David, Aja Huang, Chris J. Maddison, Arthur Guez, Laurent Sifre, George van den Driessche, Julian Schrittwieser et al. "Mastering the Game of Go with Deep Neural Networks and Tree Search." *Nature* 529, no. 7587 (January 2016): 484–89. https://doi.org/10.1038/nature16961.

Slaughter, Anne-Marie. *Unfinished Business: Women Men Work Family.* Reprint. New York: Random House Trade Paperbacks, 2016.

Smith, Aaron. *Gig Work, Online Selling and Home Sharing.* Washington, DC: Pew Research Center, 2016.

——. *Shared, Collaborative, and On Demand: The New Digital Economy.* Washington, DC: Pew Research Center, 2016.

Smith, Clint. "Wake Up, Mr. West!" *New Republic,* May 3, 2018. https://newrepublic.com/article/148222/wake-up-mr-west.

Smith, Peter. *Free-Range Learning in the Digital Age: The Emerging Revolution in College, Career, and Education.* New York: SelectBooks, 2018.

Star, Susan Leigh, and Anselm Strauss. "Layers of Silence, Arenas of Voice: The Ecology of Visible and Invisible Work." *Computer Supported Cooperative Work (CSCW)* 8, no. 1–2 (March 1, 1999): 9–30.

Stern, Andy, and Lee Kravitz. *Raising the Floor: How a Universal Basic Income Can Renew Our Economy and Rebuild the American Dream.* New York: PublicAffairs, 2016.

Stewart, Neil, Christoph Ungemach, Adam J. L. Harris, Daniel M. Bartels, Ben R. Newell, Gabriele Paolacci, and Jesse Chandler. "The Average Laboratory Samples a Population of 7,300 Amazon Mechanical Turk Workers." *Judgment and Decision Making* 10, no. 5 (2015): 13.

Stoiber, J. "Independent Contractors Should Get Benefits." *Philadelphia Inquirer,* October 20, 1996.

Stone, Brad. *The Everything Store: Jeff Bezos and the Age of Amazon.* New York: Little, Brown, 2013.

Strauss, Anselm. "The Articulation of Project Work: An Organizational Process." *Sociological Quarterly* 29, no. 2 (June 1, 1988): 163–78.

Suchman, Lucy. "Supporting Articulation Work." In *Computerization and Controversy: Value Conflicts and Social Choices,* 2nd ed., edited by Rob Kling, 407–23. San Diego, CA: Academic Press, 1996.

Sundararajan, Arun. *The Sharing Economy: The End of Employment and the Rise of Crowd-Based Capitalism.* Cambridge, MA: MIT Press, 2016.

Suri, Siddharth. "Technical Perspective: Computing with the Crowd." *Communications of the ACM* 59, no. 6 (June 2016): 101. https://doi.org/10.1145/2927926.

United Nations Development Programme. *Global Dimensions of Human Development.* New York: Oxford University Press, 1992.

U.S. Bureau of Labor Statistics. "Licensed Practical and Licensed Vocational Nurses." Occupational Outlook Handbook. Accessed June 21, 2018. https://www.bls.gov/ooh/healthcare/licensed-practical-and-licensed-vocational-nurses.htm.

U.S. Department of Labor, Wage and Hour Division. *Fact Sheet #17A: Exemption for Executive, Administrative, Professional, Computer & Outside Sales Employees Under the Fair Labor Standards Act (FLSA).* Rev. July 2008. https://www.dol.gov/whd/overtime/fs17a_overview.pdf.

U.S. Government Accountability Office. *Contingent Workforce: Size, Characteristics, Earnings, and Benefits.* GAO-15-168R. Washington, DC: Government Accountability Office, 2015.

Vaidyanathan, Geetha. "Technology Parks in a Developing Country: The Case of India." *Journal of Technology Transfer* 33, no. 3 (June 1, 2008): 285–99.

Valentine, Melissa, and Amy C. Edmondson. "Team Scaffolds: How Minimal In-Group Structures Support Fast-Paced Teaming." *Academy of Management Proceedings* 2012, no. 1 (January 1, 2012): 1. https://doi.org/10.5465/AMBPP.2012.109.

Waring, Stephen P. *Taylorism Transformed: Scientific Management Theory Since 1945.* Chapel Hill: University of North Carolina Press, 1991.

We Are Dynamo. "Dear Jeff Bezos." We Are Dynamo wiki. Accessed May 8, 2018. http://www.wearedynamo.org/dearjeffbezos.

Weber, Lauren. "Some of the World's Largest Employers No Longer Sell Things, They Rent Workers." *Wall Street Journal,* December 28, 2017. https://www.wsj.com/articles/some-of-the-worlds-largest-employers-no-longer-sell-things-they-rent-workers-1514479580.

Weil, David. *The Fissured Workplace: Why Work Became So Bad for So Many and What Can Be Done to Improve It.* Cambridge, MA: Harvard University Press, 2014.

West, Joel, and Karim R. Lakhani. "Getting Clear about Communities in Open Innovation." *Industry and Innovation* 15, no. 2 (2008): 223–31.

"What Is Jon Brelig and Oscar Smith?" *Dirtbag Requesters on Amazon Mechanical*

*Turk* (blog), August 29, 2013. http://scumbagrequester.blogspot.com/2013/08/what-is-jon-brelig-and-oscar-smith.html.

Wikipedia. S.v. "Corporate Social Responsibility." Accessed June 20, 2018. https://en.wikipedia.org/wiki/Corporate_social_responsibility.

Wikipedia. S.v. "Pareto Principle." Accessed June 15, 2018. https://en.wikipedia.org/wiki/Pareto_principle.

Williams, Joan C., Susan J. Lambert, Saravanan Kesavan; Peter J. Fugiel, Lori Ann Ospina, Erin Devorah Rapoport, Meghan Jarpe, Dylan Bellisle, Pradeep Pendem, Lisa McCorkell, and Sarah Adler-Milstein. "Stable Scheduling Increases Productivity and Sales: The Stable Scheduling Study." University of California Hastings College of the Law, University of Chicago School of Social Service Administration, University of California Kenan-Flagler Business School, 2018. https://www.ssa.uchicago.edu/stable-scheduling-study-reveals-benefits-and-feasibility-retail-families-businesses.

Wissner-Gross, Alexander. "2016: What Do You Consider the Most Interesting Recent [Scientific] News? What Makes It Important?" Edge. Accessed October 21, 2018. https://www.edge.org/response-detail/26587.

Wood, Alex. "Why the Digital Gig Economy Needs Co-Ops and Unions." openDemocracy, September 15, 2016. https://www.opendemocracy.net/alex-wood/why-digital-gig-economy-needs-co-ops-and-unions.

World Economic Forum. *Special Program of the Broadband Commission and the World Economic Forum, Meeting Report*. Geneva, Switzerland: World Economic Forum, 2018.

Yin, Ming, Mary L. Gray, Siddharth Suri, and Jennifer Wortman Vaughan. "The Communication Network Within the Crowd." In *WWW '16: Proceedings of the 25th International Conference on World Wide Web*, 1293–1303. Geneva, Switzerland: International World Wide Web Conferences Steering Committee, 2016. https://doi.org/10.1145/2872427.2883036.

Yin, Ming, Siddharth Suri, and Mary L. Gray. "Running Out of Time: The Impact and Value of Flexibility in On-Demand Crowdwork." In *CHI '18: Proceedings of the 2018 CHI Conference on Human Factors in Computing Systems*, 1–11. New York: ACM, 2018. https://doi.org/10.1145/3173574.3174004.

Zuboff, Shoshana. *In the Age of the Smart Machine: The Future of Work and Power*. New York: Basic Books, 1988.

Zyskowski, Kathryn, Meredith Ringel Morris, Jeffrey P. Bigham, Mary L. Gray, and Shaun K. Kane. "Accessible Crowdwork?: Understanding the Value in and Challenge of Microtask Employment for People with Disabilities," In *CSCW '15: Proceedings of the 18th ACM Conference on Computer Supported Cooperative Work & Social Computing*, 1682–93. New York: ACM, 2015. https://doi.org/10.1145/2675133.2675158.

# METHODS APPENDIX

Readers will find an account of the methods and data used to ground the book's quantitative claims in the following research publications, which are also cited in the endnotes.


- Chen, Wei-Chu, Mary L. Gray, and Siddharth Suri. "More than Money: Correlation Among Worker Demographics, Motivations, and Participation in Online Labor Markets." Under review, ICWSM '19: 13th International AAAI Conference on Web and Social Media, Munich, Germany, June 2019.
- Gray, Mary L., Siddharth Suri, Syed Shoaib Ali, and Deepti Kulkarni. "The Crowd Is a Collaborative Network." In *CSCW '16: Proceedings of the 19th ACM Conference on Computer-Supported Cooperative Work and Social Computing,* 134–47. New York: ACM, 2016.
- Kingsley, Sara Constance, Mary L. Gray, and Siddharth Suri. "Accounting for Market Frictions and Power Asymmetries in Online Labor Markets." *Policy & Internet* 7, no. 4 (December 1, 2015): 383–400. https://doi.org/10.1002/poi3.111.
- Yin, Ming, Mary L. Gray, Siddharth Suri, and Jennifer Wortman Vaughan. "The Communication Network Within the Crowd." In *WWW '16: Proceedings of the 25th International Conference on World Wide Web,* 1293–1303. Geneva, Switzerland: International World Wide Web Conferences Steering Committee, 2016. https://doi.org/10.1145/2872427.2883036.
- Yin, Ming, Siddharth Suri, and Mary L. Gray. "Running Out of Time: The Impact and Value of Flexibility in On-Demand Crowdwork." In *CHI '18: Proceedings of the 2018 CHI Conference on Human Factors in Computing Systems,* 1–11. New York: ACM, 2018. https://doi.org/10.1145/3173574.3174004.


Those papers focus on the surveys, experiments, and quantitative analyses conducted. Below, we provide readers with details of how we gathered the ethnographic and interview materials used to illustrate the lived experiences of ghost work featured in this book.

## ETHNOGRAPHIC FIELDWORK AND INTERVIEWS

This book draws on collaborative ethnographic fieldwork in the U.S. and in India, led by Mary L. Gray. The Ethics Advisory Board at Microsoft Research reviewed the project proposal and approved the study in January 2013. In addition to participant observation, our team conducted 189 formal, in-depth interviews (115 interviews in India and 74 interviews in the United States). and hundreds of hours of informal follow-up interviews and observations among worker research participants in their home offices or other locations where they did their work or gathered with work friends. The ethnographic observations allowed us to understand people's experiences with on-demand work and how they come to their understandings of this work and its relationship to their everyday lives.

The qualitative team in India included Shoaib Syed Ali, Mary L. Gray, and Deepti Kulkarni, conducting all field interviews and observations in India between February 10, 2013, and March 12, 2015. Mary conducted online follow-up interviews in 2017 among key India-based participants.

Following leads from our research survey responses and our mapping HIT (described in chapter 5), interviewing and fieldwork in India focused on three major IT centers—Hyderabad, Bangalore, and Chennai—as well as parts of Kerala, in the south, and Delhi, in the north. Most India-based interviews took place in people's homes, local cafés, or parks. The interviews included spending time with each participant and, in most cases, included meeting them in their homes to see their work setup and have them demonstrate how they did their work. Interviews lasted anywhere between one hour and three hours. The initial interviews included the equivalent of a $15 cash gift in appreciation for each hour individuals spent with us, recognizing that participants gave up time that could have been spent earning money doing on-demand work. Fieldwork also included observing participants in their homes, with their families and friends, and joining them at events at cricket fields, shopping bazaars, mosques, or temples that they signaled as important to them. Deepti and

Shoaib spent an average of 40 hours per week with a core group of 40 participants over the course of the India-based fieldwork, with Mary joining them for approximately six months of that time.

Mary conducted all U.S. fieldwork and interviews between February 10, 2013, and May 12, 2017, during the windows of time when she was not in India. Jason Qualls, Kristy Milland, and Kathryn Zyskowski contributed additional in-depth interviews using a similar semi-structured, open-ended interview protocol developed for the India fieldwork. U.S. fieldwork was far less extensive, involving a core group of 15 participants Mary followed over the course of the project. Observations included initial home visits to see how workers set up their workspaces and routines and observe exchanges with family members. Follow-up sessions and interviews were conducted via Skype, due to time constraints and because there was no geographical concentration of workers in the U.S., according to survey responses or the mapping experiment. U.S. workers received a $15 cash gift in appreciation for the time that individuals gave us for the initial interview. In most cases, both India and U.S. participants rejected payment for follow-up interviews.

Interview participants were recruited in the following ways: an invitation at the end of the research platform survey to participate in an in-person interview, scheduled at their convenience; worker referrals; and online contacts made on worker discussion forums. We met most fieldwork participants after they accepted the request to participate in a follow-up interview included at the end of our research survey run on the four on-demand work platforms used for this study. The remaining participants met us through snowball sampling and referral from friends and family who have done some form of on-demand work. All names used in the book are pseudonyms chosen by the research participants so that they may be able to identify themselves without compromising the identity of others in their network. Research participants included active workers, people who had tried and left on-demand work, crowdsourcing platform engineers, and entrepreneurs.

Interviews of participants in India analyzed for this paper were conducted primarily in English. These interviews were conducted in person in English by Mary, or jointly with either Shoaib or Deepti, particularly in cases where the interview participant's primary language (mother tongue) was not English. Interviews were also conducted one-on-one by Shoaib or Deepti and then reviewed with Mary at a later date. All research participants received an information sheet and an opportunity to review the intent of this study before orally consenting to participate. The research waived written consent for most participants, recognizing that the consent form itself posed the only risk to participants. A written consent document would be the only thing that linked participants to their participation in a study that, at times, criticized their sites of employment.

Lastly, in addition to worker interviews, we worked directly with an outside consulting agency, in collaboration with colleagues in Microsoft Corporate Strategy, to gather interviews that might help us understand the perspectives of those hiring on-demand workers. Though these interviews were not conducted by us, they were done in person, using the same consent process as used throughout the project, recorded, and transcribed. The consulting agency interviewed 50 full-time employees (July–October 2017), recruited from LinkedIn and other job hiring sites, representing a range of industries. Siddharth led the effort of thematically coding these interview transcripts. We use the materials to understand the perspectives and experiences of those hiring on-demand workers for what we call "macro-tasks" in the book.

# NOTES

## Introduction

1. While we are concerned that the term "ghost work" might feel unsettling or pejorative to the people who do these difficult, demanding jobs, we believe that it effectively conveys the irony at the core of this new phenomenon.
2. Aaron Smith, *Gig Work, Online Selling and Home Sharing* (Washington, DC: Pew Research Center, 2016). According to Smith, in 2016, 8 percent of U.S. adults reported earning money in the previous year by doing online tasks (such as surveys and data entry), ride hailing, shopping/delivery, cleaning/laundry, and other tasks. Kidscount.org shows census data that estimates there were 249,747,123 adults in the U.S. in 2016. Thus, 8 percent of the roughly 250 million adults gives our estimate of 20 million. The margin of error for the survey was 2.4 percent, so a more conservative estimate would be 5.6 percent of 250 million, or 14 million. Interestingly, Pew estimates that the fraction of U.S. adults who got paid to do an online task like a survey or data entry was 5 percent, whereas the fraction that got paid for ride hailing was only 2 percent.
3. James Manyika et al., *A Labor Market That Works: Connecting Talent with Opportunity in the Digital Age* (Washington, DC: McKinsey Global Institute, 2015), http://www.mckinsey.com/insights/employment_and_growth/connecting_talent_with_opportunity_in_the_digital_age.
4. John Hawksworth et al., *UK Economic Outlook: Prospects for the Housing Market and the Impact of AI on Jobs* (London: PricewaterhouseCoopers, 2017).
5. The city of Bangalore was officially changed to Bengaluru in 2014, but it is still commonly referred to as Bangalore by its local residents, including Kala.
6. Daniel W. Barowy et al., "AutoMan: A Platform for Integrating Human-Based and Digital Computation," *Communications of the ACM* 59, no. 6 (June 2016): 102–109, https://doi.org/10.1145/2927928; Siddharth Suri, "Technical Perspective: Computing with the Crowd," *Communications of the ACM* 59, no. 6 (June 2016): 101, https://doi.org/10.1145/2927926.
7. Suri, "Computing with the Crowd," 101.
8. While Ayesha is not fictional—we interviewed her in Bangalore—she did not work for CrowdFlower when we met her. She tried to sign up for Crowd-

Flower, but, after getting her account set up on MTurk with her brother's help, she focused her time on building her reputation on MTurk. We offer this as a speculative example that imagines what Ayesha's work would have looked like had she followed through on working with CrowdFlower. We hope this drives home the point that getting a sense of this labor is hard because of the constant churn among workers and the challenges of seeing and following the workers behind ghost work.

9. Erik Brynjolfsson and Andrew McAfee, *The Second Machine Age: Work, Progress, and Prosperity in a Time of Brilliant Technologies* (New York: W. W. Norton, 2014); Klaus Schwab, *The Fourth Industrial Revolution* (New York: Penguin, 2017); Erik Brynjolfsson and Andrew McAfee, *Race Against the Machine: How the Digital Revolution Is Accelerating Innovation, Driving Productivity, and Irreversibly Transforming Employment and the Economy* (Lexington, MA: Digital Frontier, 2012).
10. Tarleton Gillespie, *Custodians of the Internet: Platforms, Content Moderation, and the Hidden Decisions That Shape Social Media* (New Haven, CT: Yale University Press, 2018), 18–19.
11. See Frederick Daso, "Bill Gates and Elon Musk Are Worried for Automation —But This Robotics Company Founder Embraces It," *Forbes*, December 18, 2017, https://www.forbes.com/sites/frederickdaso/2017/12/18/bill-gates-elon-musk-are-worried-about-automation-but-this-robotics-company-founder-embraces-it/; Jasper Hamill, "Elon Musk's Fears of AI Destroying Humanity Are 'Speciesist', Said Google Boss," *Metro* (blog), May 2, 2018, https://metro.co.uk/2018/05/02/elon-musks-fears-artificial-intelligence-will-destroy-humanity-speciesist-according-google-founder-larry-page-7515207/; "Stephen Hawking: 'I fear AI may replace humans altogether' The theoretical physicist, cosmologist and author talks Donald Trump, tech monopolies and humanity's future," *Wired*, November 28, 2017, https://www.wired.co.uk/article/stephen-hawking-interview-alien-life-climate-change-donald-trump.
12. See, for example, "Robots? Is Your Job at Risk?," CNN, September 15, 2017; "When the Robots Take Over, Will There Be Jobs Left for Us?," CBS News, April 9, 2017; "More Robots, Fewer Jobs," Bloomberg, May 8, 2017.
13. Alex Ross, *The Industries of the Future* (New York: Simon and Schuster, 2016); Stephen A. Herzenberg, John A. Alic, and Howard Wial, *New Rules for a New Economy: Employment and Opportunity in Post-Industrial America* (Ithaca, NY: ILR Press, 2000); Chris Brenner, *Work in the New Economy: Flexible Labor Markets in Silicon Valley*, Information Age Series (Malden, MA: Wiley-Blackwell, 2002).
14. Scott Hartley, *The Fuzzy and the Techie: Why the Liberal Arts Will Rule the Digital World* (Boston: Houghton Mifflin Harcourt, 2017). Hartley focuses on the case of AlphaGo. Both AlphaGo and AlphaGo Zero were the brainchildren of DeepMind, a London-based research lab acquired by Google in 2014.
15. Tom Dietterich, personal conversation, April 13, 2018. Noted AI researcher

Dietterich put it this way: the version of AlphaGo that defeated Ke Jie was "told" the rules of go (in the sense that it could invoke code to compute all legal moves for any board state and it was given the definitions of winning and losing). It was also given a large database of games between human experts. That database was used with supervised learning to train the initial move selection function (the "policy function") for AlphaGo. Then AlphaGo carried out a second phase of "self play" in which it played against a copy of itself (a technique first developed by Arthur Samuel in 1959, I believe) and applied reinforcement learning algorithms to refine the policy function. Finally, they ran additional self-play games to learn a "value function" (value network) that predicted which side would win in each board state. During game play, AlphaGo combines the value function with the policy function to select moves based on forward search (Monte Carlo tree search).

16. David Silver et al., "Mastering the Game of Go with Deep Neural Networks and Tree Search," *Nature* 529, no. 7587 (January 2016): 484–89, https://doi.org/10.1038/nature16961.
17. For a more theoretically rich account of the mixing of human effort and computational processes, see H. R. Ekbia and Bonnie A. Nardi, *Heteromation, and Other Stories of Computing and Capitalism* (Cambridge, MA: MIT Press, 2017).
18. Bureau of Labor Statistics, "Contingent and Alternative Employment Arrangements, May 2017," Economic News Release, U.S. Department of Labor, June 7, 2018.
19. U.S. Government Accountability Office, *Contingent Workforce: Size, Characteristics, Earnings, and Benefits*, GAO-15-168R (Washington, DC: Government Accountability Office, 2015).
20. Lawrence F. Katz and Alan B. Krueger, "The Rise and Nature of Alternative Work Arrangements in the United States, 1995–2015" (NBER Working Paper Series, no. 22667, National Bureau of Economic Research, Cambridge, MA, September 2016).
21. Diana Farrell and Fiona Greig, *The Online Platform Economy: Has Growth Peaked?* (JPMorgan Chase Institute, 2017).
22. David Weil, *The Fissured Workplace: Why Work Became So Bad for So Many and What Can Be Done to Improve It* (Cambridge, MA: Harvard University Press, 2014).
23. This research would not have been possible without Wei-Chu Chen, Sara Kingsley, Greg Minton, and Ming Yin, who contributed to the original quantitative data analyses, and Syed Shoaib Ali, Deepti Kulkarni, Jason Qualls, and Kathryn Zyskowski, who contributed to the book's qualitative data analyses. Mary led all fieldwork and worker interviews conducted in the United States and India, the survey design, historical analyses found in chapter 2, and qualitative analyses throughout the book. Siddharth led all online experiments and quantitative analyses found through the book, as well as the qualitative analyses of interviews with fellow requesters found in chapter 3. Invaluable

research assistance was also provided along the way by Andrea Alarcón, Sarah Hamid, Rebecca Hoffman, Kate Miltner, Christopher Persaud, and Steven Schirra.

24. Winter Mason and Siddharth Suri, "Conducting Behavioral Research on Amazon's Mechanical Turk," *Behavior Research Methods* 44, no. 1 (March 2012): 1–23, https://doi.org/10.3758/s13428-011-0124-6.

## 1. Humans in the Loop

1. M. Six Silberman, "Human-Centered Computing and the Future of Work: Lessons from Mechanical Turk and Turkopticon, 2008–2015" (PhD diss., University of California, Irvine, 2015).
2. Brad Stone, *The Everything Store: Jeff Bezos and the Age of Amazon* (New York: Little, Brown, 2013).
3. Ibid.
4. Ibid.
5. It's rumored that MTurk was a pet project of Jeff Bezos himself, who saw it as a tool that could not only serve MTurk internally but generate income as a new marketplace serving its sellers who also needed help cleaning up their listings. See Stone, *The Everything Store.*
6. Daniel W. Barowy et al., "VoxPL: Programming with the Wisdom of the Crowd," in *CHI '17: Proceedings of the 2017 CHI Conference on Human Factors in Computing* Systems (New York: ACM, 2017), 2347–58, https://doi.org/10.1145/3025453; Suri, "Computing with the Crowd," 101.
7. Much has been written critiquing Amazon's dismissive naming of a work platform, though it's not clear that the company ever planned to make the platform something publicly visible. It has officially been in beta since its launch in 2005.
8. Kevin P. Murphy, *Machine Learning: A Probabilistic Perspective* (Cambridge, MA: MIT Press, 2012).
9. Fei-Fei Li, "ImageNet: Where Have We Been? Where Are We Going?," ACM Learning Webinar, https://learning.am.org/, accessed September 21, 2017; Deng et al., "ImageNet: A Large-Scale Hierarchical Image Database," in *2009 IEEE Conference on Computer Vision and Pattern Recognition* (Piscataway, NJ: IEEE), 248–55.
10. Accuracy went from 72 percent to 97 percent between 2010 and 2016.
11. In fact, Alexander Wissner-Gross said, "Data sets — not algorithms — might be the limiting factor to development of human-level artificial intelligence." See Alexander Wissner-Gross, "2016: What Do You Consider the Most Interesting Recent [Scientific] News? What Makes It Important?" Edge, https://www.edge.org/response-detail/26587, accessed October 21, 2018.
12. To incentivize researchers to use the data set, Li and her colleagues organized an annual contest pitting the best algorithms for the image recognition problem, from various research teams around the world, against one another.

The progress scientists made toward this goal was staggering. The annual ImageNet competition saw a roughly 10x reduction in error and a roughly 3x increase in precision in recognizing images over the course of eight years. Eventually the vision algorithms achieved a lower error rate than the human workers. The algorithmic and engineering advances that scientists achieved over the eight years of competition fueled much of the recent success of neural networks, the so-called deep learning revolution, which would impact a variety of fields and problem domains.

13. Djellel Difallah, Elena Filatova, and Panos Ipeirotis, "Demographics and Dynamics of Mechanical Turk Workers," in *Proceedings of the Eleventh ACM International Conference on Web Search and Data Mining* (New York: ACM, 2018), 135–43, https://doi.org/10.1145/3159652.3159661.
14. Ibid.
15. Panos Ipeirotis, "How Many Mechanical Turk Workers Are There?," *A Computer Scientist in Business School* (blog), January 29, 2018, http://www.behind-the-enemy-lines.com/.
16. We intentionally asked workers to self-report their location so that they could report their location down to any level of granularity they were comfortable sharing. Since the map allowed workers to search, zoom, and pan as they saw fit, they were free to put a pin on their house, neighborhood, county, city, etc. After placing a pin on their location and clicking "save," workers were then shown a Bing map of the world with the pins of the last 500 workers to do this HIT. Each pin was randomly moved a little to protect worker privacy, and they were reminded of that on this page. This HIT could easily be done in less than one minute, and we paid 25 cents for completing it.
17. Hara et al., "A Data-Driven Analysis of Workers' Earnings on Amazon Mechanical Turk," 2018 CHI Conference on Human Factors in Computing Systems, Paper No. 449, 2018.
18. Tools like the web browser extension Turkopticon help workers share information about requesters through a review and rating system, grading requesters on their communication, pricing, fairness, and prompt payment or response time to worker queries. Workers also have elaborate, self-moderated forums for swapping more details behind paywalls. Neither requester reputations nor worker forums are available directly on MTurk. See Lilly C. Irani and M. Six Silberman, "Turkopticon: Interrupting Worker Invisibility in Amazon Mechanical Turk," in *CHI '13: Proceedings of the SIGCHI Conference on Human Factors in Computing Systems* (New York: ACM, 2013).
19. See Siou Chew Kuek, Cecilia Paradi-Guilford, Toks Fayomi, Saori Imaizumi, Panos Ipeirotis, Patricia Pina, and Manpreet Singh, "The Global Opportunity in Online Outsourcing," World Bank Group, June 2015, and Panos Ipeirotis, "How Big Is Mechanical Turk?," *A Computer Scientist in Business School* (blog), November 8, 2012, http://www.behind-the-enemy-lines.com/2012/11/is-mechanical-turk-10-billion-dollar.html.

20. Amazon charges 20 percent extra for HITs with more than ten assignments. Each assignment corresponds to a unique worker.
21. Ironically, as employees of Microsoft, we are under the same NDAs as those doing ghost work, limiting what we can share about the specifics of the UHRS platform's operations.
22. Annalee Newitz, "The Secret Lives of Google Raters," *Ars Technica*, April 27, 2017, https://arstechnica.com/features/2017/04/the-secret-lives-of-google-raters/.
23. Gillespie, *Custodians of the Internet;* Jeremias Prassl, *Humans as a Service: The Promise and Perils of Work in the Gig Economy* (Oxford, England: Oxford University Press, 2018); Sarah T. Roberts, "Digital Detritus: 'Error' and the Logic of Opacity in Social Media Content Moderation," *First Monday* 23, no. 3 (March 1, 2018), http://firstmonday.org/ojs/index.php/fm/article/view/8283; Sarah T. Roberts, "Social Media's Silent Filter," *The Atlantic*, March 8, 2017, https://www.theatlantic.com/technology/archive/2017/03/commercial-content-moderation/518796/.
24. In the 1960s, India became a big player in outsourcing for information and communication services, financial services, and much of what we call today "back office" work or "business processes." As more and more processes like auditing and the documentation of accounting purchases moved to centralized databases accessed through secured servers and away from paper and in-office management and storage of files, fewer and fewer people needed to be employed on-site, full-time, to file and shuffle records and other key business documents.
25. Sarah T. Roberts, "Commercial Content Moderation: Digital Laborers' Dirty Work," in S. U. Noble and B. Tynes, eds., *The Intersectional Internet: Race, Sex, Class and Culture Online* (New York: Peter Lang, 2016), 147–59. See also Sarah T. Roberts, *Behind the Screen: Content Moderation in the Shadows of Social Media* (New Haven, CT: Yale University Press, forthcoming).
26. LeadGenius began by winning business contracts with U.S. cities looking for an inexpensive way to keep their municipal websites updated. Changes to information about where to send parking ticket payments or instructions on how residents could get a new curb cut for their driveway were easy to outsource. LeadGenius (then called MobileWorks) relied on workers in India as well as thousands of American workers recruited from state-sponsored career advancement programs for people on public assistance.
27. Our surveys of LeadGenius focus on India and the U.S., but it is important to note that the company has teams of what they call researchers — workers helping them generate leads for LeadGenius clients — in 40 countries. So some of our findings differ slightly from the demographics that LeadGenius publicly reports.
28. In June 2016, LeadGenius stopped hiring data researchers and managers in the following countries: Australia, Barbados, Canada, France, Germany, Greece,

Hong Kong, Ireland, Italy, Jamaica, Japan, Mauritius, Oman, Poland, Portugal, Russia (Moscow), Saint Kitts and Nevis, Saudi Arabia, Singapore, Slovenia, Spain, St. Martin, the UK, and the USA. But LeadGenius is hiring in Afghanistan, Albania, Algeria, Angola, Argentina, Armenia, Azerbaijan, Bangladesh, Barbados, Belarus, Belize, Benin, Bhutan, Bolivia, Bosnia and Herzegovina, Botswana, Brazil, Bulgaria, Burkina Faso, Burundi, Cambodia, Cameroon, Cape Verde, the Central African Republic, Chad, Colombia, Comoros, Costa Rica, Côte d'Ivoire, Croatia, Cuba, the Czech Republic, the Democratic Republic of Congo, Djibouti, Dominica, the Dominican Republic, Ecuador, Egypt, El Salvador, Eritrea, Ethiopia, Fiji, Gabon, Gambia, Georgia, Ghana (West Africa), Grenada, Guatemala, Guinea, Guinea-Bissau, Guyana, Haiti, Honduras, India, Indonesia, Iran, Iraq, Jordan, Kazakhstan, Kenya, Kiribati, Kosovo, Kyrgyzstan, Laos, Latvia, Lebanon, Libya, Lithuania, Macedonia, Madagascar, Malawi, Malaysia, Maldives, Mali, the Marshall Islands, Mauritania, Mexico, Micronesia, Moldova, Mongolia, Montenegro, Morocco, Mozambique, Myanmar, Namibia, Nepal, Nicaragua, Nigeria, Pakistan, Palau, Palestine, Panama, Papua New Guinea, Paraguay, Peru, the Philippines, the Republic of Congo, Romania, Russia (outside Moscow), Rwanda, Samoa, São Tomé and Principe, Senegal, Serbia, Sierra Leone, the Solomon Islands, Somalia, South Africa, South Sudan, Sri Lanka, St. Lucia, St. Vincent and the Grenadines, Sudan, Suriname, Swaziland, Syria, Taiwan, Tajikistan, Tanzania, Thailand, Timor-Leste, Togo, Tonga, Trinidad and Tobago, Tunisia, Turkey, Tuvalu, the UAE, Ukraine, Uzbekistan, Vanuatu, Venezuela, Vietnam, the West Bank, Yemen, Zambia, and Zimbabwe.

29. Together, Jansen, Reville, Chang, and Wilson created the open-source project called Miro that seeded Amara's vision. At its height, Miro had about two million users a month — not quite a revolution, but not too bad a following, either. A few for-profit ventures started around the same time. One was a small startup soon acquired by Google. It was called YouTube. YouTube quickly eclipsed Miro, leaving PCF unsure of what it could do with the open-source software it had poured years into building to share videos on the web. By 2010 PCF was still focused on Miro, but it also had some side projects simmering. One of them emerged out of a conversation between Jansen and Wilson. Wilson's wife, originally from Brazil, wanted to share some Portuguese films with friends. They were trying to find subtitles for these films, and when they couldn't, they repurposed some of the principles behind Miro to create a web-based software-editing kit that could make it easy to add subtitles to a streaming video. Eventually their prototype became the first version of Amara, a web-based interface for linking video to captions.

People could use the Amara online editing tools to watch a television show in Spanish and add English subtitling, making the content available in other languages and more accessible to those with hearing challenges. PCF continued to refine Amara, adding features like wiki-style collaboration options to share notes about the videos. PCF eventually won awards for its software —

one from the Federal Communications Commission, for accessibility, another from the UN, for intercultural connection. They had momentum for mainstreaming the captioning tools. All they needed was a lot of video content and an even larger international network of volunteers interested in using their tools. And then a group of volunteer translators who were using other tools to translate TED Talks asked for a chance to use Amara instead.

30. According to the nonprofit organization's website, the first six TED Talks—highly produced presentations that converge around technology, entertainment, and design—were posted online on June 27, 2006. By September they had reached more than one million views. TED Talks proved so popular that in 2007, TED's website was relaunched around them, giving a global audience free access to some of the world's greatest thinkers, leaders, and teachers. TED's Open Translation Project (OTP) launched in 2009, with 300 translations in 40 languages, created by 200 volunteer translators. Today, more than 120,000 translations have been published, in 115 languages (and counting), created by more than 28,000 volunteers. In 2012, the program expanded to include the transcription and translation of TEDx Talks, the translation of TED-Ed lessons, and the translation of content distributed by worldwide partners who help grow TED's global footprint.
31. Amara, by definition of its mission, is an international network. Close to 70 percent of our survey respondents live outside the United States. As such, the details of our survey results include AOD members beyond the U.S. and India.
32. See Wei-Chu Chen, Mary L. Gray, and Siddharth Suri, "More than Money: Correlation Among Worker Demographics, Motivations, and Participation in Online Labor Markets," under review, ICWSM '19: The 13th International AAAI Conference on Web and Social Media, Munich, Germany, June 2019.
33. See Prassl, *Humans as a Service*.

## 2. From Piecework to Outsourcing

1. Erin Hatton, *The Temp Economy: From Kelly Girls to Permatemps in Postwar America* (Philadelphia: Temple University Press, 2011). See also Louis Hyman, *Temp: How American Work, American Business, and the American Dream Became Temporary* (New York: Penguin, 2018).
2. As Yale historian David Blight has discussed, the four million slaves forced to labor in the U.S. in 1860 were "worth more than every bank, factory, and railroad in the country combined." See Clint Smith, "Wake Up, Mr. West!," *New Republic,* May 3, 2018, https://newrepublic.com/article/148222/wake-up-mr-west.
3. On the ontology of whiteness, historian Nell Irvin Painter writes, "Constructions of whiteness have changed over time, shifting to accommodate the demands of social change. Before the mid-19th century, the existence of more than one white race was commonly accepted, in popular culture and scholar-

ship. Indeed, there were several. Many people in the United States were seen as white — and could vote (if they were adult white men) — but were nonetheless classified as inferior (or superior) white races. Irish-Americans present one example." See Nell Irvin Painter, "What Is Whiteness?," *New York Times,* December 21, 2017, https://www.nytimes.com/2015/06/21/opinion/sunday/what-is-whiteness.html; Nell Irvin Painter, *The History of White People,* reprint (New York: W. W. Norton, 2011).

4. In fact, anti-abolitionist and worker movements clashed as young immigrant, predominantly Irish, men railed against freed slaves moving north as "the new army of the unemployed," threatening to unmoor their shaky footing as wage workers. If not for the U.S. government's series of so-called Homestead Acts, from 1862 through the 1930s, which distributed land bought or seized from indigenous peoples to move swelling urban populations westward, factory towns' wages may have bottomed out and ethnic wars between immigrant groups would have been far bloodier. See David R. Roediger, *The Wages of Whiteness: Race and the Making of the American Working Class* (London: Verso, 1999).
5. Harnessing electricity and Thomas Newcomen's invention of the first commercial steam-powered engine, in 1712, made twice as powerful by James Watt in 1781, propelled manufacturing forward.
6. Stephen Waring argues that Frederick Taylor's principles of scientific management focused much of early-day industry on wrestling control from the tradesmen coordinating most skilled craftwork. As Waring puts it, "New managerial capitalism emerged from a search for ways to coordinate operations and control workers." Stephen P. Waring, *Taylorism Transformed: Scientific Management Theory Since 1945* (Chapel Hill: University of North Carolina Press, 1991). Also see Shelley Pennington and Belinda Westover, "Types of Homework," in *A Hidden Workforce,* Women in Society (London: Palgrave Macmillan, 1989), 44–65, https://doi.org/10.1007/978-1-349-19854-2_4.
7. "Homework" was formally defined as "the manufacture or preparation within the home of goods intended for sale, in which the work supplements the factory process." Pennington and Westover, "Types of Homework."
8. Marjorie Abel and Nancy Folbre further elaborate the ways in which women's piecework was neglected by formal accounts. At the beginning of the century, the majority of families depended on one another, particularly wives, mothers, and daughters, to bake, can, cook, clean, and sew what was needed. All members of the family played some part in the division of labor that kept everyone fed, healthy, and clothed. Women directed and led tasks that involved working by hand and with machines to spin, card, weave, and knit together raw fibers and turn them into pants, shirts, sweaters, and work clothes for the day. A family was both a social safety net and a small industrial unit. As cities grew, bakeshops, laundries, and large garment factories cropped up. These became the first consumable goods and services humanity had ever had beyond the

home, an extension of domestic ovens, washtubs, and sewing kits. Less than three generations before, only royalty and slave owners could afford to purchase someone else to feed, house, and clothe themselves. But as these services became industries in their own right, beyond private consumption at home, women followed. See Nancy Folbre, "Women's Informal Market Work in Massachusetts, 1875–1920," *Social Science History* 17, no. 1 (1993): 135–60, https://doi.org/10.2307/1171247, and Marjorie Abel and Nancy Folbre, "A Methodology for Revising Estimates: Female Market Participation in the U.S. Before 1940," *Historical Methods: A Journal of Quantitative and Interdisciplinary History* 23, no. 4 (October 1, 1990): 167–76, https://doi.org/10.1080/01615440.1990.10594207.

9. Elizabeth Beardsley Butler joined the Russell Sage Foundation to do the first and what would turn out to be one of the only comprehensive studies of the expendable labor pools of pieceworkers working in the shadow of industrialization as it rises as the dominant form of employment. See Elizabeth Beardsley Butler, *Women and the Trades* (Pittsburgh: Charities Publication Committee, 1909).
10. Ibid., 13.
11. Ibid., 23.
12. "Early economic historians saw industrial home work as a transitory stage between artisan production and the factory system and assumed that as industrialization got under way, such forms of labor would disappear. This, however, has not been the case and industrial home work has continued to exist in some form or another through the industrialization process." Sandra Albrecht, "Industrial Home Work in the United States: Historical Dimensions and Contemporary Perspective," *Economic and Industrial Democracy* 3, no. 4 (1982): 414, https://doi.org/10.1177/0143831X8234003.
13. Butler, "Women and the Trades," 139.
14. Ibid., 139.
15. Ibid., 134.
16. David Noble, *Forces of Production: A Social History of Industrial Automation* (New York: Routledge, 2017).
17. Hatton, *The Temp Economy.*
18. This was 3.5 percent of the total U.S. population in 1946, and at the time around 23 percent of working Americans were in manufacturing sectors.
19. See John Barnard, *Walter Reuther and the Rise of the Auto Workers* (Boston: Little, Brown, 1983); Eldorous Dayton, *Walter Reuther: The Autocrat of the Bargaining Table* (New York: Devin-Adain, 1958); Victor Reuther, *The Brothers Reuther and the Story of UAW* (Boston: Houghton Mifflin, 1976); Nelson Lichtenstein, *The Most Dangerous Man in Detroit* (New York: Basic Books, 1995).
20. The act also ended employees' rights to walk out on wildcat strikes without authorization from union leadership and required unions to give employers 80 days' notice of a pending strike, in many cases triggering a further require-

ment that unions sit down to arbitration before they exercised their right to strike. Strikes related to political candidates or held to disrupt another business's production were no longer protected. And, in one of the biggest blows to union growth, Taft-Hartley made unionizing a states' rights issue. Individual states were now allowed to pass right-to-work laws that banned requiring union dues at workplaces organized by unions.

21. From the mid-19th to the early 20th century, paperboys (or newsies) were tasked with hawking newspapers for the many daily outlets that competed for readers. This was a common first job for adolescents in Western nations, most often boys, where the expectation was to work very long hours to avoid being penalized for unsold papers. See M. Schuman, "History of Child Labor in the United States—Part 1: Little Children Working," *Monthly Labor Review,* U.S. Bureau of Labor Statistics, January 2017, https://www.bls.gov/opub/mlr/2017/article/history-of-child-labor-in-the-united-states-part-1.htm; *National Labor Relations Board v. Hearst Publications,* 322 U.S. 111 (1944).
22. See *National Labor Relations Board,* 322 U.S. 111 (1944); Weil, *The Fissured Workplace,* 185–86.
23. The FLSA requires that most employees in the United States be paid at least the federal minimum wage for all hours worked and overtime pay at time and a half the regular rate of pay for all hours worked over 40 hours in a workweek. However, Section 13(a)(1) of the FLSA provides an exemption from both minimum wage and overtime pay for employees employed as bona fide executive, administrative, professional, and outside sales employees. Section 13(a)(1) and Section 13(a)(17) also exempt certain computer employees. To qualify for exemption, employees generally must meet certain tests regarding their job duties and be paid on a salary basis at not less than $455 per week. Job titles do not determine exempt status. In order for an exemption to apply, an employee's specific job duties and salary must meet all the requirements of the Department of Labor's regulations. See U.S. Department of Labor, Wage and Hour Division, *Fact Sheet #17A: Exemption for Executive, Administrative, Professional, Computer & Outside Sales Employees Under the Fair Labor Standards Act (FLSA),* rev. July 2008, https://www.dol.gov/whd/overtime/fs17a_overview.pdf.
24. Jonathan Grossman, "Fair Labor Standards Act of 1938: Maximum Struggle for a Minimum Wage," Office of the Assistant Secretary for Administration and Management, U.S. Department of Labor website. Originally published in *Monthly Labor Review,* June 1978, https://www.dol.gov/oasam/programs/history/flsa1938.htm.
25. Young women were hired on contract to operate secure lines for the Coast Guard and Navy yards in the area so that they could call from ship to land. They were all let go after the war and none were paid benefits or severance of any kind. See Jill Frahm, "The Hello Girls: Women Telephone Operators with the American Expeditionary Forces During World War I," *Journal of the Gilded Age and Progressive Era* 3, no. 3 (2004): 271–93.

26. Some of the source material for this definition is no longer available on the U.S. Department of Labor blog. However, a cached version of a blog post obtained through the Wayback Machine, written by former labor secretary Tom Perez and former commerce secretary Penny Pritzker, serves as a good example of the prevailing ethos for much of the Obama administration that there is a skills gap that needs to be addressed for middle-class people to succeed. However, this is distinct from "middle skill jobs," which are defined as those that require more education and training than secondary education but less than a four-year university degree. Allegedly, there is a high demand for workers who can fill these middle-skill jobs, but too many unemployed and underemployed Americans do not have the training to take them on. This issue has been alternatively framed as a "talent supply chain problem," a "skills shortage," and a "crisis of the middle skilled workforce," among others, by business leaders and policy makers in the past couple of decades. See "Middle Skills," U.S. Competitiveness, Harvard Business School website, accessed May 22, 2018, https://www.hbs.edu/competitiveness/research/Pages/middle-skills.aspx.; Francis Green, *Skills and Skilled Work: An Economic and Social Analysis* (Oxford, England: Oxford University Press, 2013); Tom Perez and Penny Pritzker, "A Joint Imperative to Strengthen Skills," *The Commerce Blog,* U.S. Department of Commerce, September 11, 2013; and Peter Smith, *Free-Range Learning in the Digital Age: The Emerging Revolution in College, Career, and Education* (New York: SelectBooks, 2018).
27. See Jennifer Light, "When Computers Were Women," *Technology and Culture* 40, no. 3 (July 1999), 455–83; Greg Downey, "Virtual Webs, Physical Technologies, and Hidden Workers," *Technology and Culture* 42, no. 2 (April 2001): 209–35; David Allen Grier, *When Computers Were Human* (Princeton, NJ: Princeton University Press, 2005).
28. Langley Field became NASA Langley Research Center, aka NASA Langley and LaRC, in 1958.
29. Nathalia Holt, *Rise of the Rocket Girls: The Women Who Propelled Us, from Missiles to the Moon to Mars,* reprint (New York: Back Bay, 2017).
30. Margot Lee Shetterly, *Hidden Figures: The American Dream and the Untold Story of the Black Women Mathematicians Who Helped Win the Space Race,* media tie-in ed. (New York: William Morrow, 2016), 4.
31. Ibid.
32. Ibid., 21.
33. Ibid., 61.
34. Holt, *Rocket Girls.*
35. Ibid.
36. Ibid.
37. Hatton, *The Temp Economy.*
38. Geetha Vaidyanathan, "Technology Parks in a Developing Country: The Case of India," *Journal of Technology Transfer* 33, no. 3 (June 1, 2008): 285–99;

Dinesh C. Sharma, *The Outsourcer: The Story of India's IT Revolution* (Cambridge, MA: MIT Press, 2015).

39. Weil, *The Fissured Workplace.*
40. Hatton, *The Temp Economy;* S. Greenhouse, "Equal Work, Less-Equal Perks: Microsoft Leads the Way in Filling Jobs With 'Permatemps,'" *New York Times,* March 30, 1998; J. Stoiber, "Independent Contractors Should Get Benefits," *Philadelphia Inquirer,* October 20, 1996.
41. Levy and Murnane's notions of "expert thinking" and "complex communication" are a good way to understand the technical reasons for the division of labor between computers and humans. For Levy and Murnane, expert thinking requires "solving new problems for which there are no routine solutions," and complex communication involves "persuading, explaining, and in other ways conveying a particular interpretation of information." The other tasks carried out by U.S. labor forces are "routine cognitive tasks," like maintaining expense reports and other services that are "well described by logical rules"; "routine manual tasks," like "physical tasks that can be well described by using rules," such as "installing windshields"; and "non-routine manual tasks that cannot be well described as following a set of If-Then-Do rules because they require optical recognition and fine motor control" – "like driving a truck." "This drive to develop, produce, and market new products relies on the human ability to manage and solve analytical problems and communicate new information, so it keeps expert thinking and complex communication in strong demand." See Frank Levy and Richard Murnane, *The New Division of Labor: How Computers Are Creating the Next Job Market* (Princeton, NJ: Princeton University Press, 2004).
42. Shoshana Zuboff, *In the Age of the Smart Machine: The Future of Work and Power* (New York: Basic Books, 1988).
43. Also called "piece-rate," "putting-out," British "cottage industries," "industrial home work," and the American "commission system." See Albrecht, "Industrial Home Work," 413–30.
44. Lauren Weber, "Some of the World's Largest Employers No Longer Sell Things, They Rent Workers," *Wall Street Journal,* December 28, 2017, https://www.wsj.com/articles/some-of-the-worlds-largest-employers-no-longer-sell-things-they-rent-workers-1514479580.
45. S&P study cited in Weber, "Some of the World's Largest Employers." The breadth of services on offer from outsourcing firms is staggering. Compass Group was founded in 1941 to run factory cafeterias in wartime England, eventually branching out into corporate catering. It now employs more than 550,000 and counts among its subsidiaries firms like Eurest Services, which staffs and manages mailrooms for clients, provides them with full-time receptionists, sets up their conference rooms for meetings, and operates their warehouses. Eurest's clients include Google, S&P, and Pfizer. Outsourcing firms' workforces may shrink as algorithms take on more tasks, says Steve Hall, a

partner at Information Services Group Inc. (ISG). "The large outsourcers are using a combination of analytics and automation to significantly reduce the need for labor," he says.

46. According to research and advisory firm ISG.

## 3. Algorithmic Cruelty and the Hidden Costs of Ghost Work

1. Eric Meyer, "Inadvertent Algorithmic Cruelty," *Meyerweb* (blog), December 24, 2014, https://meyerweb.com/eric/thoughts/2014/12/24/inadvertent-algorithmic-cruelty/; revised version published on Slate.com on December 29, 2014. Meyer's post was a response to the rollout of Facebook's "Year in Review" feature. In December, an unwelcome post appeared on Meyer's Facebook feed—a close-cropped picture of Meyer's five-year-old daughter, Rebecca, who had died of brain cancer earlier that year. The header addressed him directly: "Eric, here's what your year looked like!" Practically speaking, how could Facebook predict who might enjoy the serendipity of these memories and who might experience them as unnecessary and unwelcome reminders of a painful loss? Facebook's unilateral decision to pluck photos from Meyer's timeline and place them in his newsfeed rotation seems coldhearted. What cruelty is this? But Meyer did not assume cruel intentions. He underscored the banality behind a commercial platform rolling out a feature that it could assume would generate more "stickiness" on the platform for the majority of its users. It chose to release this new product feature "at scale" with a heavy-handed approach that included no "off switch" for the pictorial flashbacks and, initially, no way to make the images private to prevent others from "liking" the photo of a lost child. With access to 93 million users' photos and posts, Facebook has plenty of material to curate (see Gillespie, "Politics of 'Platforms.'"). What is less clear is Facebook's latitude to repurpose users' memories and impose them as a new feature without explicit guidance from users. But the sheer scale of managing its collective user base makes weighing individual needs not just unrealistic but seemingly impossible. Nonetheless, users can be left feeling like the site doesn't care about them (at best) or is exploitative or callous (at worst).
2. Our intent is not to downplay the emotional distress of seeing a deceased child's face pop up in a Facebook app meant to showcase the past year's highlights. Lee Humphreys also discusses Facebook's algorithmic curation of memory within her theory of media accounting. See Lee Humphreys, *The Qualified Self: Social Media and the Accounting of Everyday Life* (Cambridge, MA: MIT Press, 2018), 85–90.
3. R. H. Coase, "The Nature of the Firm," *Economica* 4, no. 16 (1937): 388, https://doi.org/10.1111/j.1468-0335.1937.tb00002.x.
4. Mason and Suri, "Amazon's Mechanical Turk," 1–23.
5. For generative, pivotal critiques of this framing, see Ilana Gershon, *Down*

*and Out in the New Economy: How People Find (or Don't Find) Work Today* (Chicago: University of Chicago Press, 2017); Melissa Gregg, *Work's Intimacy* (Cambridge, England: Polity, 2011); Melissa Gregg, *Counterproductive: Time Management in the Knowledge Economy* (Durham, NC: Duke University Press, 2018); Gina Neff, *Venture Labor: Work and the Burden of Risk in Innovative Industries* (Cambridge, MA: MIT Press, 2012); Trebor Scholz, *Uberworked and Underpaid: How Workers Are Disrupting the Digital Economy* (Cambridge, England: Polity, 2016).

6. Our results show that many workers share lucrative tasks and information about reputable requesters with their network connections. With access to this extra information, connected workers might be able to start on high-quality tasks before other workers hear about them. In the extreme, this might lead to connected workers using up all of the high-paying tasks before isolated workers have had a chance to find them, effectively starving out the isolated workers. Thus, we speculate that being a part of the network may confer an advantage to workers. See Ming Yin et al., "The Communication Network Within the Crowd," in *WWW '16: Proceedings of the 25th International Conference on World Wide Web* (Geneva, Switzerland: International World Wide Web Conferences Steering Committee, 2016), 1293–1303, https://doi.org/10.1145/2872427.2883036.
7. Workers were randomly assigned to each of these treatments. See Ming Yin, Siddharth Suri, and Mary L. Gray, "Running Out of Time: The Impact and Value of Flexibility in On-Demand Crowdwork," in *CHI '18: Proceedings of the 2018 CHI Conference on Human Factors in Computing Systems* (New York: ACM, 2018), 1–11, https://doi.org/10.1145/3173574.3174004.
8. Ibid. Workers were randomly assigned to one of these six treatments (two durations times three pay rates).
9. Sara Horowitz, "Special Report: The Costs of Nonpayment," *Freelancers Union Blog*, accessed May 8, 2018, http://blog.freelancersunion.org/2015/12/10/costs-nonpayment/.
10. Steven Hill, *How (Not) to Regulate Disruptive Business Models* (Berlin: Friedrich Ebert Stiftung, 2016).
11. Cathy O'Neil, *Weapons of Math Destruction: How Big Data Increases Inequality and Threatens Democracy* (New York: Crown, 2016), 21.
12. Panagiotis G. Ipeirotis, "Analyzing the Amazon Mechanical Turk Marketplace," *XRDS: Crossroads, The ACM Magazine for Students* 17, no. 2 (December 1, 2010): 16, https://doi.org/10.1145/1869086.1869094.
13. Sara Constance Kingsley, Mary L. Gray, and Siddharth Suri, "Accounting for Market Frictions and Power Asymmetries in Online Labor Markets," *Policy & Internet* 7, no. 4 (December 1, 2015): 383–400, https://doi.org/10.1002/poi3.111. See also Arindrajit Dube et al., "Monopsony in Online Labor Markets," *American Economic Review: Insights* (forthcoming).
14. John Horton, "Online Labor Markets," in *Internet and Network Economics: 6th*

*International Workshop, WINE 2010, Stanford, CA, USA, December 13–17, 2010, Proceedings,* Lecture Notes in Computer Science (New York: Springer, 2011). Horton suggests, and we agree, that "the influence of the market creator is so pervasive that their role in the market is closer to that of a government . . . they determine the space of permissible actions within the market, such as what contractual forms are allowed and who is allocated decision rights."

15. Juliet B. Schor and Craig J. Thompson, *Sustainable Lifestyles and the Quest for Plenitude: Case Studies of the New Economy* (New Haven, CT: Yale University Press, 2014).

## 4. Working Hard for (More Than) the Money

1. Board of Governors of the Federal Reserve System, *Report on the Economic Well-Being of U.S. Households in 2016* (Washington, DC: Federal Reserve Board, May 2017); Neal Gabler, "The Secret Shame of Middle-Class Americans," *The Atlantic,* May 2016, https://www.theatlantic.com/magazine/archive/2016/05/my-secret-shame/476415/.
2. Brooke Erin Duffy calls this mix of passion projects and portfolio-building forms of "aspirational labor." See Brooke Erin Duffy, *(Not) Getting Paid to Do What You Love: Gender, Social Media, and Aspirational Work* (New Haven, CT: Yale University Press, 2017).
3. Weil, *The Fissured Workplace;* Dean Baker, *Rigged: How Globalization and the Rules of the Modern Economy Were Structured to Make the Rich Richer* (Washington, DC: Center for Economic and Policy Research, 2016); Herzenberg, Alic, and Wial, *New Rules.*
4. Weil, *The Fissured Workplace.*
5. Arlie Russell Hochschild, *The Managed Heart: Commercialization of Human Feeling,* 3rd ed. (Berkeley: University of California Press, 2012).
6. Neff, *Venture Labor;* Ursula Huws, *Labor in the Global Digital Economy: The Cybertariat Comes of Age* (New York: Monthly Review Press, 2014); Alice E. Marwick, *Status Update: Celebrity, Publicity, and Branding in the Social Media Age* (New Haven, CT: Yale University Press, 2013).
7. Jay Shambaugh Nantz et al., *Thirteen Facts About Wage Growth* (Washington, DC: Brookings Institution, September 25, 2017), https://www.brookings.edu/research/thirteen-facts-about-wage-growth/.
8. Lawrence Mishel, Elise Gould, and Josh Bivens, *Wage Stagnation in Nine Charts* (Washington, DC: Economic Policy Institute, 2015), http://www.epi.org/publication/charting-wage-stagnation/.
9. National Low Income Housing Coalition, "Out of Reach" (Washington, DC: National Low Income Housing Coalition, 2018), http://nlihc.org/oor; Susan J. Lambert, Peter J. Fugiel, and Julia R. Henly, "Precarious Work Schedules Among Early-Career Employees in the US: A National Snapshot," research brief (Chicago: EINet, University of Chicago, 2014), 24; Dan Clawson and

Naomi Gerstel, *Unequal Time: Gender, Class, and Family in Employment Schedules* (New York: Russell Sage Foundation, 2014); Bridget Ansel and Heather Boushey, *Modernizing U.S. Labor Standards for 21st-Century Families,* The Hamilton Project (Washington, DC: Brookings Institution, 2017), 25; Lydia DePillis, "The Next Labor Fight Is Over When You Work, Not How Much You Make," *Wonkblog, Washington Post,* May 8, 2015, https://www.washingtonpost.com/news/wonk/wp/2015/05/08/the-next-labor-fight-is-over-when-you-work-not-how-much-you-make/; Robert Reich, "How the New Flexible Economy Is Making Workers' Lives Hell," *Robert Reich* (blog), April 20, 2015, http://robertreich.org/post/116924386855.

10. Board of Governors of the Federal Reserve, *Economic Well-Being.*
11. Several cities (San Francisco, Seattle, and New York, to name a few) and the state of Oregon have passed legislation on scheduling reforms and protections for employees who are usually subjected to just-in-time schedules. The recent boom of scheduling and management software businesses, like Celayix, Smartsheet, and Shiftboard, that repurpose ghost work's tool kit of APIs and websites offer so-called predictive scheduling, suggesting that the practice of automating employees' schedules is far from dying out. See Joan C. Williams et al., "Stable Scheduling Increases Productivity and Sales: The Stable Scheduling Study," University of California Hastings College of the Law, University of Chicago School of Social Service Administration, University of California Kenan-Flagler Business School, 2018, https://www.ssa.uchicago.edu/stable-scheduling-study-reveals-benefits-and-feasibility-retail-families-businesses.
12. Chen, Gray, and Suri, "More than Money."
13. Mary L. Gray et al., "The Crowd Is a Collaborative Network," in *CSCW '16: Proceedings of the 19th ACM Conference on Computer-Supported Cooperative Work & Social Computing* (New York: ACM, 2016), 134–47, https://doi.org/10.1145/2818048.2819942.
14. Ibid.
15. Ahmed et al., "Peer-to-Peer in the Workplace: A View from the Road," in *CHI: '16. Proceedings of the 2016 CHI Conference on Human Factors in Computing Systems* (New York: ACM, 2016), 5063–75, https://doi.org/10.1145/2858036.2858393.
16. Julie Yujie Chen, "Thrown Under the Bus and Outrunning It! The Logic of Didi and Taxi Drivers' Labour and Activism in the On-Demand Economy," *New Media & Society*, September 6, 2017, https://doi.org/10.1177/1461444817729149.
17. Wikipedia, s.v. "Pareto Principle," accessed June 15, 2018, https://en.wikipedia.org/wiki/Pareto_principle.
18. United Nations Development Programme, *Global Dimensions of Human Development,* Human Development Report (New York: Oxford University Press, 1992).
19. Jesse Chandler, Pam Mueller, and Gabriele Paolacci, "Nonnaïveté Among Amazon Mechanical Turk Workers: Consequences and Solutions for Behav-

ioral Researchers," *Behavior Research Methods* 46, no. 1 (March 2014): 112–30, https://doi.org/10.3758/s13428-013-0365-7.

20. Stewart et al., "The Average Laboratory Samples a Population of 7,300 Amazon Mechanical Turk Workers," *Judgment and Decision Making* 10, no. 5 (2015): 13; Karën Fort, Gilles Adda, and K. Bretonnel Cohen, "Amazon Mechanical Turk: Gold Mine or Coal Mine?," *Computational Linguistics* 37, no. 2 (2011): 413–20.
21. Ruth Schwartz Cowan, *More Work for Mother: The Ironies of Household Technology from the Open Hearth to the Microwave*, 2nd ed. (New York: Basic Books, 1985).
22. Arlie Hochschild and Anne Machung, *The Second Shift: Working Families and the Revolution at Home*, rev. ed (New York: Penguin, 2012); Gregg, *Work's Intimacy*.
23. Winifred R. Poster, "Hidden Sides of the Credit Economy: Emotions, Outsourcing, and Indian Call Centers," *International Journal of Comparative Sociology* 54, no. 3 (June 2013): 205–27, https://doi.org/10.1177/0020715213501823.
24. Anne-Marie Slaughter, *Unfinished Business: Women Men Work Family*, repr. ed. (New York: Random House Trade Paperbacks, 2016).
25. LinkedIn posted business-card-transcription tasks through an intermediary using the requester name "Oscar Smith." See M. Six Silberman and Lilly Irani, "Operating an Employer Reputation System: Lessons from Turkopticon, 2008–2015," *Comparative Labor Law & Policy Journal*, February 8, 2016, http://papers.ssrn.com/abstract=2729498. MTurk workers flagged "Oscar Smith" on forums and community blogs as someone who was notorious for poor compensation. See "What Is Jon Brelig and Oscar Smith?," *Dirtbag Requesters on Amazon Mechanical Turk* (blog), August 29, 2013, http://scumbagrequester.blogspot.com/2013/08/what-is-jon-brelig-and-oscar-smith.html, for workers' comments like "The copy business cards under the guise of Oscar Smith is for an app called CardMunch which is owned by LinkedIn. Yes, a multi-million dollar corporation is the one behind paying workers less than $1 per hour . . . New non-usa accounts have not been accepted on Mturk for close to a year. They are gradually eliminating these foreign accounts because of low quality work. And yes, Indians and new turkers are primarily the ones working for LinkedIn aka Oscar Smith and Jon Brelig because not a single American with even a shred of self esteem or self worth would work for these scumbags."
26. Though workers in the U.S. may have less costly or easier access to computers and an internet connection, that does not mean they have the financial resources to make earning money a lower priority. See Chen, Gray, and Suri, "More than Money."
27. Ibid.
28. Ibid.
29. Ibid.
30. Duffy, *(Not) Getting Paid*.
31. Chen, Gray, and Suri, "More than Money."

32. For stories from Americans in various gig work arrangements, see John Bowe, Marisa Bowe, and Sabin Streeter, eds., *Gig: Americans Talk About Their Jobs,* (New York: Broadway Books, 2001). For more on shadow economies and informal employment, see LaShawn Harris, *Sex Workers, Psychics, and Numbers Runners: Black Women in New York City's Underground Economy* (Urbana: University of Illinois Press, 2016).
33. Hochschild and Machung, *The Second Shift;* Gregg, *Work's Intimacy;* Kylie Jarrett, *Feminism, Labour and Digital Media: The Digital Housewife* (New York: Routledge, 2015).
34. Sareeta Amrute, *Encoding Race, Encoding Class: Indian IT Workers in Berlin,* reprint (Durham, NC: Duke University Press, 2016).
35. "The Rights of Persons with Disabilities Act, 2016 (the "Disabilities Act, 2016") along with the Rights of Persons with Disabilities Rules, 2017 (together, the "Disability Law") has been enacted by the Indian government.

    "The new Disability Law gives effect to the principles of the *United Nations Convention on the Rights of Persons with Disabilities.* The Disability Law *inter alia* seeks to protect disabled persons from various forms of discrimination, increases measures for effective participation and inclusion in the society, and ensures equality of opportunity and adequate accessibility.

    "Prior to enactment of the Disabilities Act, 2016, the law governing rights of the disabled were scattered across the Constitution of India, the Persons with Disabilities (Equal Opportunity Protection of Rights and Full Participation) Act, 1995 ("Disabilities Act, 1995"), the Mental Health Act, 1987, the Rehabilitation Council Act of India and the National Trust (for welfare of persons with Autism, Cerebral Palsy, Metal [*sic*] Retardation and Multiple Disabilities) Act, 1999. Although these legislations aimed at safeguarding the rights of persons with disabilities, these legislations did not specifically provide for *equality of opportunity* especially in matters relating to employment.

    "The Disabilities Act, 2016 has repealed the Disabilities Act, 1995." See http://www.disabilityaffairs.gov.in/.
36. Shehzad Nadeem, *Dead Ringers: How Outsourcing Is Changing the Way Indians Understand Themselves,* reprint (Princeton, NJ: Princeton University Press, 2013), Simon Denyer, *Rogue Elephant: Harnessing the Power of India's Unruly Democracy* (New York: Bloomsbury Press, 2014).
37. Broadband Commission for Sustainable Development, *State of Broadband 2017: Broadband Catalyzing Sustainable Development* (Geneva, Switzerland: Broadband Commission for Sustainable Development, 2017); World Economic Forum, *Special Program of the Broadband Commission and the World Economic Forum, Meeting Report* (Geneva, Switzerland: World Economic Forum, 2018).
38. *Special Program of the Broadband Commission and the World Economic Forum Meeting Report,* World Economic Forum, Davos, Switzerland, 2018.
39. Labor economists would say that requesters use contingent labor to find the worker who'll do the job for the lowest pay. More liberal economists argue that

people's being able to use their marketable skills and bargain for higher wages makes flexibility a good thing for workers. You'd have to believe that (1) there's a pure market setting wages in opposition to collusion among requesters — like what happens in Silicon Valley to keep a cap on engineers' salaries — and (2) wages reflect a true market value of what people contribute (so janitors, most on flexible contract, not salaried, earn so little because they do work that's worth less than that of the engineers whose offices they clean). See Weil, *The Fissured Workplace*; Baker, *Rigged*.

## 5. The Kindness of Strangers and the Power of Collaboration

1. Sociologist Anselm Strauss called this *articulation work,* or the "overall process of putting *all* the work elements together *and* keeping them together." See Anselm Strauss, "The Articulation of Project Work: An Organizational Process," *Sociological Quarterly* 29, no. 2 (June 1, 1988): 163–78.
2. Much as Susan Leigh Star anticipated, computer adoption, and with it assumptions that computational processes would take over the cognitive load of collaboration, made articulation work even more central, as it "manages the consequences of the distributed nature of the work" that are otherwise thought solved by technologies. See Susan Leigh Star and Anselm Strauss, "Layers of Silence, Arenas of Voice: The Ecology of Visible and Invisible Work," *Computer Supported Cooperative Work (CSCW)* 8, no. 1–2 (March 1, 1999): 9–30.
3. Anthropologist Lucy Suchman, building on Strauss, uses the elegant phrase "artful integrations" to describe articulation work as "aspects of systems development and use that have been hidden, or at least positioned in the background or shadows, and to bring them forward into the light." See Lucy Suchman, "Supporting Articulation Work," in *Computerization and Controversy,* ed. Rob Kling (San Diego, CA: Academic Press, 1996), 407–23.
4. Gray et al., "The Crowd," 134–47.
5. Google AdWords was renamed Google Ads in 2018.
6. Workers were asked to create a nickname for themselves to preserve their anonymity, and then we asked them to swap nicknames with any workers they communicate with. They could use any medium they wished to swap nicknames. The task ran from August to September 2016. For more details, see Yin et al., "The Communication Network."
7. There are more than 265,000 business processing jobs in Bangalore, India, alone. Most who work these jobs have never been to America or Great Britain; their cultural knowledge is exclusive to what they've picked up in school, in training, through movies, and in routine conversations with customers. Accordingly, a cottage industry has developed, offering private courses that claim to help boost English proficiency, conversational skills, and Western cultural mannerisms. The primary goal of these courses is accent neutralization, or eradicating what is known as mother tongue influence (MTI) — the traces

of accent and vocal inflections that remain after learning English as a second language. The goal isn't just to sound like an American; companies like Orion Edutech, for instance, also attend to cultural nuance — teaching Western customs by showing American TV shows like *Friends* or *Seinfeld* and exposing Indians to American popular music. See A. Aneesh, *Neutral Accent: How Language, Labor, and Life Become Global* (Durham, NC: Duke University Press, 2015); J. K. Tina Basi, *Women, Identity and India's Call Centre Industry* (London: Routledge, 2009); Mahuya Pal and Patrice Buzzanell, "The Indian Call Center Experience: A Case Study in Changing Discourses of Identity, Identification, and Career in a Global Context," *Journal of Business Communication* 45, no. 1 (January 1, 2008): 31–60, https://doi.org/10.1177/0021943607309348; Sumita Raghuram, "Identities on Call: Impact of Impression Management on Indian Call Center Agents," *Human Relations* 66, no. 11 (November 1, 2013): 1471–96, https://doi.org/10.1177/0018726713481069.

8. A quote from interviews with TurkerNation discussion forum users indicates that some workers value online forums for both reasons: "If I had not found TurkerNation, I would not have made as much money for sure. And the fun we have when things are slow: priceless."
9. Yin et al.,"The Communication Network."
10. Ibid.
11. Ibid.
12. Ibid.
13. The data from workers reporting where they are is shown in the maps in figures 1A and 1B. Here we're more interested in how they found out about the HIT. For more details on how we gathered this data, see Kingsley, Gray, and Suri, "Market Frictions," 383–400, and Gray et al., "The Crowd," 134–47.
14. See Salehi et al., "We Are Dynamo: Overcoming Stalling and Friction in Collective Action for Crowd Workers," in *CHI '15: Proceedings of the 33rd Annual ACM Conference on Human Factors in Computing Systems* (New York: ACM, 2015), 1621–30, https://doi.org/10.1145/2702123.2702508.
15. We Are Dynamo, "Dear Jeff Bezos," We Are Dynamo wiki, accessed May 8, 2018, http://www.wearedynamo.org/dearjeffbezos.

## 6. The Double Bottom Line

1. By comparison, it took three years for national relief efforts to distribute even 12 percent of international funds raised for reconstruction, and it was limited to helping those who owned their own property, further deepening social fault lines. Almost 75 percent of the money from the company's Earthquake Relief Fund went directly to CloudFactory's workers and their families, and the rest was donated to local government agencies and nonprofits offering food and shelter around the city.
2. Thomas Fox Parry, "The Death of a Gig Worker," *The Atlantic*, June 1, 2018,

https://www.theatlantic.com/technology/archive/2018/06/gig-economy-death/561302/?utm_source=atltw.

3. See *National Labor Relations Board,* 322 U.S. 111 (1944); Weil, *The Fissured Workplace,* 185–86.
4. Aaron Smith, *Shared, Collaborative, and On Demand: The New Digital Economy* (Washington, DC: Pew Research Center, 2016), http://www.pewinternet.org/2016/05/19/on-demand-ride-hailing-apps/.
5. Left unsettled is whether all of Uber's driver-partners can be considered a category of workers eligible to bring a class-action lawsuit against Uber or whether driver-partners will need to decide what to do as individuals to protect their private interests. See Mike Isaac and Noam Scheiber, "Uber Settles Cases with Concessions, but Drivers Stay Freelancers," *New York Times,* April 21, 2016, http://www.nytimes.com/2016/04/22/technology/uber-settles-cases-with-concessions-but-drivers-stay-freelancers.html; Alex Rosenblat and Luke Stark, "Algorithmic Labor and Information Asymmetries: A Case Study of Uber's Drivers," *International Journal of Communication* 10 (July 27, 2016): 27.
6. Alex Rosenblat, *Uberland: How Algorithms Are Rewriting the Rules of Work* (Oakland: University of California Press, 2018).
7. Wikipedia, s.v. "Corporate Social Responsibility," accessed June 20, 2018, https://en.wikipedia.org/wiki/Corporate_social_responsibility.
8. Gray et al., "The Crowd," 134–47.
9. Parikh's approach to corporate social responsibility took things one step further. He asked company founders to design products and services that made a market for themselves, filling a societal need rather than using profits from a popular product to fund philanthropy. Parikh had some powerful fans, like the venture capitalists behind the Bay Area incubator Y Combinator. Students who enrolled in Parikh's class had a chance to compete for real financial backing through Y Combinator. Philipp Gutheim, Anand Kulkarni, Prayag Narula, and Dave Rolnitzky took their classroom project, MobileWorks, and won Y Combinator's summer 2011 competition, giving them enough money to bankroll a group of engineers, a marketing campaign, and a network of on-demand workers serving as virtual assistants from around the world. See Gray et al., "The Crowd"; Anand Kulkarni et al., "MobileWorks: Designing for Quality in a Managed Crowdsourcing Architecture," *IEEE Internet Computing* 16, no. 5 (September 2012): 28–35, https://doi.org/10.1109/MIC.2012.72.
10. Joel West and Karim R. Lakhani, "Getting Clear about Communities in Open Innovation," *Industry and Innovation* 15, no. 2 (2008): 223–31; Boudreau et al., "From Crowds to Collaborators: Initiating Effort & Catalyzing Interactions Among Online Creative Workers," HBS Working Paper No. 14-060 (Cambridge, MA: Harvard Business School, January 2014).
11. As Narula notes, "Self-organization doesn't just happen . . . The crowd itself needs help, support, and the way to do that is building some sort of hierarchi-

cal system within the crowd where more experienced people get to move out and move into more management roles where they're helping other people, where they're providing support, where they're directly interfacing with their clients and they're representing the community. We are all of that in this organization." Prayag Narula, field research interview, December 11, 2015.

12. The organization officially changed its name from Universal Subtitles to Amara in 2012. The group wanted a name that would allow it to add other services, beyond subtitles, and it wanted to capture the sense of community that it hoped to build: a nonprofit with a social mission, rather than a typical web startup. The word Amara was chosen because it is a form of the Spanish verb *amar,* "to love," and it also means "eternal" in Sanskrit. Co-founder Dean Jansen also noted in our interviews with him that Amara's "lawyers were also concerned about potential trademark issues with 'Universal Films / Studios,' since they had been in the media space for so long."
13. Arun Sundararajan, *The Sharing Economy: The End of Employment and the Rise of Crowd-Based Capitalism* (Cambridge, MA: MIT Press, 2016); Airi Lampinen et al., "Studying the 'Sharing Economy': Perspectives to Peer-to-Peer Exchange," in *Proceedings of the 18th ACM Conference Companion on Computer Supported Cooperative Work & Social Computing* (New York: ACM, 2015), 117–21, https://doi.org/10.1145/2685553.2699339; Juliet Schor, "Debating the Sharing Economy," Great Transition Initiative, October 2014, http://www.greattransition.org/publication/debating-the-sharing-economy; Schor et al., "Paradoxes of Openness and Distinction in the Sharing Economy," *Poetics* 54 (2016): 66–81.
14. Kristofer Erickson and Inge Sørensen, "Regulating the Sharing Economy," *Internet Policy Review* 5, no. 3 (June 30, 2016), https://doi.org/10.14763/2016.2.414; Juho Hamari, Mimmi Sjöklint, and Antti Ukkonen, "The Sharing Economy: Why People Participate in Collaborative Consumption," *Journal of the Association for Information Science and Technology,* 2015; Aaron Smith, *Shared, Collaborative, and On Demand.*
15. "We believe the new economy is creating opportunities to reinvent work, but we need to ensure the end goal is work that is good for workers." National Domestic Workers Alliance, "The Good Work Code for the Online Economy Announces First 12 Companies Leading for Good for Workers," press release, November 13, 2015, via Marketwired, http://www.marketwired.com/press-release/good-work-code-online-economy-announces-first-12-companies-leading-good-work-workers-2073469.htm.
16. Trebor Scholz, "Platform Cooperativism vs. the Sharing Economy," *Trebor Scholz* (blog), December 5, 2014, https://medium.com/@trebors/platform-cooperativism-vs-the-sharing-economy-2ea737f1b5ad; Alex Wood, "Why the Digital Gig Economy Needs Co-Ops and Unions," openDemocracy, September 15, 2016, https://www.opendemocracy.net/alex-wood/why-digital-gig-economy-needs-co-ops-and-unions; Chelsea Rustrum, "Q&A with Felix

Weth of Fairmondo, the Platform Co-Op That's Taking on eBay," Shareable, accessed June 21, 2018, https://www.shareable.net/blog/qa-with-felix-weth-of-fairmondo-the-platform-co-op-thats-taking-on-ebay; Nithin Coca, "Nurses Join Forces with Labor Union to Launch Healthcare Platform Cooperative," Shareable, accessed June 21, 2018, https://www.shareable.net/blog/nurses-join-forces-with-labor-union-to-launch-healthcare-platform-cooperative.

17. U.S. Bureau of Labor Statistics, "Licensed Practical and Licensed Vocational Nurses," Occupational Outlook Handbook, accessed June 21, 2018, https://www.bls.gov/ooh/healthcare/licensed-practical-and-licensed-vocational-nurses.htm.

18. John Bellamy Foster, Robert W. McChesney, and R. Jamil Jonna, "The Global Reserve Army of Labor and the New Imperialism," *Monthly Review* 63, no. 6 (2011): 1. See also Mark Graham, Isis Hjorth, and Vili Lehdonvirta, "Digital Labour and Development: Impacts of Global Digital Labour Platforms and the Gig Economy on Worker Livelihoods," *Transfer: European Review of Labour and Research* 23, no. 2 (2017): 135–62, https://doi.org/10.1177/1024258916687250.

## Conclusion

1. See Siou Chew Kuek, Cecilia Paradi-Guilford, Toks Fayomi, Saori Imaizumi, Panos Ipeirotis, Patricia Pina, and Manpreet Singh, "The Global Opportunity in Online Outsourcing," World Bank Group, June 2015; for related, in some cases more conservative, estimates, see Lawrence Mishel, *Uber and the Labor Market,* Washington, DC: Economic Policy Institute, 2018, https://www.epi.org/publication/uber-and-the-labor-market-uber-drivers-compensation-wages-and-the-scale-of-uber-and-the-gig-economy/; James Manyika et al., *Independent Work: Choice, Necessity, and the Gig Economy* (Washington, DC: McKinsey Global Institute: October 2016), http://www.mckinsey.com/global-themes/employment-and-growth/independent-work-choice-necessity-and-the-gig-economy; James Manyika et al., *Harnessing Automation for a Future That Works* (Washington, DC: McKinsey Global Institute: January 2017), http://www.mckinsey.com/global-themes/digital-disruption/harnessing-automation-for-a-future-that-works; Till Alexander Leopold, Saadia Zahidi, and Vesselina Ratcheva, *The Future of Jobs: Employment, Skills and Workforce Strategy for the Fourth Industrial Revolution,* World Economic Forum, 2016.

2. According to Smith, *Gig Work,* in 2016, 8 percent of U.S. adults reported earning money in the previous year by doing online tasks (such as surveys and data entry), ride hailing, shopping/delivery, cleaning/laundry, and other tasks. Kidscount.org shows census data that estimates there were 249,747,123 adults in the U.S. in 2016. Thus, 8 percent of the roughly 250 million adults gives our estimate of 20 million. The margin of error for the survey was 2.4 percent, so a more conservative estimate would be 5.6 percent of 250 million, or 14 million.

3. See new on-demand services targeted at enterprise clients, such as Upwork Enterprise, Hubstaff, Outsourcely, and PeoplePerHour.
4. Manyika et al. *Independent Work.*
5. Projection based on the following: "Overall, we estimate that 50 percent of the activities that people are paid to do in the global economy have the potential to be automated by adapting currently demonstrated technology. While less than 5 percent of occupations can be fully automated, about 60 percent have at least 30 percent of activities that can technically be automated" by 2055. Manyika et al., *Harnessing Automation.*
6. Kingsley, Gray, and Suri. "Accounting for Market Frictions," 383–400.
7. These are actually two of Siddharth's MTurk worker IDs.
8. David H. Autor, "Why Are There Still So Many Jobs? The History and Future of Workplace Automation," *Journal of Economic Perspectives* 29, no. 3 (Summer 2015): 3–30; Brynjolfsson and McAfee, *The Second Machine Age.*
9. For a more detailed analysis of how people with disabilities take up on-demand work, see Kathryn Zyskowski, Meredith Ringel Morris, Jeffrey P. Bigham, Mary L. Gray, and Shaun K. Kane, "Accessible Crowdwork?: Understanding the Value in and Challenge of Microtask Employment for People with Disabilities," in *CSCW '15: Proceedings of the 18th ACM Conference on Computer Supported Cooperative Work & Social Computing,* 1682–93 (New York: ACM, 2015), https://doi.org/10.1145/2675133.2675158.
10. Melissa Valentine and Amy C. Edmondson, "Team Scaffolds: How Minimal In-Group Structures Support Fast-Paced Teaming," *Academy of Management Proceedings* 2012, no. 1 (January 1, 2012): 1, https://doi.org/10.5465/AMBPP.2012.109; Dietmar Harhoff and Karim R. Lakhani, eds., *Revolutionizing Innovation: Users, Communities, and Open Innovation* (Cambridge, MA: MIT Press, 2016).
11. Coase, "Nature of the Firm," 386–405.
12. Katz and Krueger, "Alternative Work Arrangements."
13. National Domestic Workers Alliance website, accessed June 21, 2018, https://www.domesticworkers.org/.
14. Andy Stern and Lee Kravitz, *Raising the Floor: How a Universal Basic Income Can Renew Our Economy and Rebuild the American Dream* (New York: PublicAffairs, 2016); Alyssa Battistoni, "The False Promise of Universal Basic Income," *Dissent,* Spring 2017, https://www.dissentmagazine.org/article/false-promise-universal-basic-income-andy-stern-ruger-bregman; Rana Foroohar, "We're About to Live in a World of Economic Hunger Games," *Time,* July 19, 2016, http://time.com/4412410/andy-stern-universal-basic-income/; Thomas Piketty, *Capital in the Twenty-First Century,* trans. Arthur Goldhammer, reprint (Cambridge, MA: Belknap Press of Harvard University Press, 2017).
15. "Common Ground for Independent Workers," *From the WTF? Economy to the Next Economy* (blog), November 10, 2015. https://wtfeconomy.com/common-ground-for-independent-workers-83f3fbcf548f#.ey89fvtnn.

# INDEX